PROCESS TECHNOLOGY FOR SILICON CARBIDE DEVICES

Edited by

Carl-Mikael Zetterling

KTH, Royal Institute of Technology, Sweden

Published by: INSPEC, The Institution of Electrical Engineers, London, United Kingdom

© 2002: The Institution of Electrical Engineers

British Library Cataloguing in Publication Data

Process technology for silicon carbide devices.–(EMIS processing series; no. 2)
1. Silicon carbide—Electric properties
I. Zetterling, C. M. II. Institution of Electrical Engineers
621.3′8152

ISBN 0 85296 998 8

Printed in England by Antony Rowe Ltd

EMIS Processing Series No. 2

Series Advisor: Professor B. L. Weiss

PROCESS TECHNOLOGY FOR SILICON CARBIDE DEVICES

ELECTRONIC MATERIALS INFORMATION SERVICE

This comprises two series of books from INSPEC:

EMIS Processing series

The present volume is the second in the EMIS Processing series, which is designed to complement the EMIS Datareviews series.

Details of No.1 Silicon Wafer Bonding Technology for VLSI and MEMS applications (2002) is available at www.iee.org.uk/Publish/Books/Emisp/

EMIS Datareviews series

Details of the following books and links to online ordering facilities for both series can be found at www.iee.org.uk/Publish/Books/Emis/

Contents

Contents

2 Bulk and epitaxial growth of SiC 13
N. Nordell

Contents

Contents

Contents

Contents

Editor

C.-M. Zetterling
Department of Microelectronics and Information Technology
KTH, Royal Institute of Technology, SE-16440 Kista, Sweden

Carl-Mikael Zetterling received the M.Sc.E.E. and Ph.D. degrees from KTH, the Royal Institute of Technology in Stockholm, in 1991 and 1997, respectively. He is currently associate professor in the Department of Microelectronics and Information Technology. He is director of studies at the department, and teaches VLSI process technology and modelling of VLSI devices. His field of research is process technology and device design of high-frequency and high-power devices in wide bandgap materials, especially SiC. He has been working experimentally on SiC since 1992, and has contributed to over 60 publications. He was a visiting scholar for one year at the Center for Integrated Systems (CIS) at Stanford University, USA, and has also been to Japan as visiting professor (Kyoto University in 1998 and Kyoto Institute of Technology in 2001).

Authors

S.-M. Koo
Department of Microelectronics and Information Technology
KTH, Royal Institute of Technology, SE-16440 Kista, Sweden

Sang-Mo Koo received his B.S. and M.S. degrees in electrical engineering from Korea University, Seoul, Korea, in 1997 and 1999. In 1999, he spent three months as a visiting researcher at Francis Bitter Magnet Lab, MIT, studying magnetic memory junctions. In 2000 he joined the Department of Microelectronics and Information Technology, KTH, Royal Institute of Technology, Stockholm, Sweden, where he is currently pursuing a Ph.D. degree on field-effect transistors as well as metal-ferroelectric-semiconductor devices in SiC. He has contributed to about 20 scientific papers.

S.-K. Lee
Department of Microelectronics and Information Technology
KTH, Royal Institute of Technology, SE-16440 Kista, Sweden

Sang-Kwon Lee received his B.S. degree in physics from Dongguk University, Seoul, Korea, in 1988 and M.S. degree in applied physics from the University of Texas at Arlington, USA, in 1991. From 1994 to 1998 he was a researcher at Mando R&D Center in Korea, where he was involved in developing various sensors and actuators for automotive applications. In 2002, he received his Ph.D. degree from KTH, the Royal Institute of Technology, Stockholm, Sweden. He is the author or co-author of around 25 technical papers and has extensive experience in various processes in SiC contacts and devices.

N. Nordell
KTH Semiconductor Laboratory
Electrum 229, SE-16440 Kista, Sweden

Nils Nordell received his M.S. and Ph.D. degrees from KTH, the Royal Institute of Technology, Stockholm, Sweden, in 1987 and 1995, respectively. He worked with MOVPE technology for growth of III–V materials for optoelectronic devices at the

Institute of Microelectronics, Stockholm, from 1987 to 1993. In 1993 he studied SiC growth at Case Western Reserve University, Cleveland, OH, USA, and between 1993 and 1998 he was responsible for the development of SiC epitaxial growth technology for electronic devices at the Industrial Microelectronics Center in Stockholm. He is currently director of the KTH Semiconductor Laboratory.

M. Östling
Department of Microelectronics and Information Technology
KTH, Royal Institute of Technology, SE-16440 Kista, Sweden

Mikael Östling received the M.S. degree in engineering physics in 1980 and the Ph.D. degree in electronics in 1983, both from Uppsala University. In 1984 Östling joined the faculty at the School of Electrical Engineering, Royal Institute of Technology (KTH), Stockholm, where he holds a position as professor and head of the Department of Microelectronics and Information Technology. His field of research is process and device technology for Si/SiGe and SiC/wide bandgap materials in high-frequency and high-power applications. His research comprises more than 250 scientific papers, four book chapters and several invited talks.

S.J. Pearton
Department of Materials Science and Engineering
University of Florida, Gainesville, FL 32611, USA

Steve Pearton received his Ph.D. in physics from the University of Tasmania in 1983. He was a research officer at the Australian Atomic Energy Commission from 1982 to 1983, a postdoctoral researcher at Lawrence Berkeley Laboratory and UC Berkeley from 1983 to 1984 and a Member of Technical Staff at AT&T Bell Laboratories from 1984 to 1994. He has published over 800 journal publications and edited or written 10 books. His research interests are in semiconductor materials and devices.

A. Schöner
Acreo AB, Department of Micro Technology
Electrum 236, SE-16440 Kista, Sweden

Adolf Schöner graduated in physics at the Institute of Applied Physics, University of Erlangen-Nürnberg, Germany, and received his Ph.D. in 1994. He joined the Industrial Microelectronics Center (IMC) in Kista, Sweden, to continue working with material and processing issues of SiC, specialising in electrical characterisation of SiC epitaxial layers. At Acreo AB he is now responsible for the SiC epitaxial material development, as well as the design and installation of specialised measurement equipment for electrical characterisation of material, processing techniques, and devices.

E.Ö. Sveinbjörnsson
Department of Microelectronics
Chalmers University of Technology, SE-41296 Göteborg, Sweden

Einar Ö. Sveinbjörnsson is an Associate Professor at the Department of Microelectronics at Chalmers University of Technology, Göteborg, Sweden. His main research objectives are characterisation and optimisation of dielectrics on SiC and their use in high-frequency MOSFET devices. Einar received his B.Sc. in physics at the University of Iceland, M.S.E.E. in electrical engineering at the University of Southern California, Los Angeles, and Ph.D. at Chalmers University of Technology in 1994. He spent the years of 1995–1996 as a research associate at the Max-Planck Institute for Solid State Research in Stuttgart, Germany.

Abbreviations

Acronym	Chapter	Meaning
ACCUFET	7	accumulation mode MOSFET
AFM	6	atomic force microscopy
BJT	7	bipolar junction transistor
CV	3, 5, 6	capacitance-voltage measurement
CVD	2, 6	chemical vapour deposition
DDLTS	3	double correlated DLTS
DIMOSFET	7	double-implanted MOSFET
DLTS	3, 5	deep level transient spectroscopy
DMOSFET	7	double-diffused MOSFET
ECR	4	electron cyclotron resonance
EPR	3	electron paramagnetic resonance
ESR	3	electron spin resonance
FE	6	field emission
FET	7	field-effect transistor
GSG	7	ground-signal-ground
GTO	7	gate turn-off (thyristor)
HBT	7	heterojunction bipolar transistor
HEMT	7	high electron mobility transistor
HF	5, 7	hydrofluoric acid
HTCVD	2	high temperature chemical vapour deposition
ICP	4, 7	inductively coupled plasma
IGBT	7	insulated gate bipolar transistor
IMPATT	7	impact ionization avalanche transit-time
IV	6	current-voltage measurement
JBS	7	junction barrier Schottky
JFET	7	junction field-effect transistor
JTE	7	junction termination extension
JVD	5	jet vapour deposition

LDMOSFET	7	lateral double-diffused MOSFET
LED	1, 7	light emitting diode
LEO	7	lateral epitaxial overgrowth
LPE	2	liquid phase epitaxy
MEMS	7	micro-electro-mechanical system
MESFET	7	metal-semiconductor field-effect transistor
MISFET	5	metal-insulator-semiconductor field-effect transistor
MODFET	7	modulation doped field-effect transistor
MOSFET	5, 7	metal-oxide-semiconductor field-effect transistor
MQW	7	multiple quantum well
NO	5	nitrous oxide
NPT	1, 7	non punch-through
ONO	5	oxide-nitride-oxide
PBT	7	permeable base transistor
PEC	4, 7	photoelectrochemical etching
PMA	5	post-metallization annealing
PT	1, 7	punch-through
RBS	3, 6	Rutherford backscattering spectrometry
RF	2, 3	radio frequency
RIE	4	reactive ion etching
RPECVD	5	remote plasma enhanced chemical vapour deposition
RTP	3	rapid thermal processing
SAW	7	surface acoustic waves
SEM	6	scanning electron microscopy
SEMOSFET	7	static channel expansion MOSFET
SIAFET	7	static induction injected accumulated FET
SIMS	3, 6	secondary ion mass spectrometry
SIT	7	static induction transistor
TE	6	thermionic emission
TEM	3, 6	transmission electron microscopy
TFE	6	thermionic field emission
TLM	6	transfer length method or transmission line method
TSC	5	thermally stimulated current

Abbreviations

UMOSFET	5, 7	U-shaped groove MOSFET
UV	1, 4	ultraviolet
VJFET	7	vertical JFET
VLSI	7	very large scale integration
VPE	2	vapour phase epitaxy
XPS	6	X-ray photoelectron spectroscopy
XRD	6	X-ray diffraction

Introduction

Silicon carbide (SiC) has been known as a semiconductor material almost longer than silicon, but it is still only thought of as a semiconductor for niche applications. There are several reasons for this, one being the problem of making SiC material of sufficient quality. In the late 1980s SiC wafers became available from a commercial vendor, which opened the possibility for many groups to experiment with electronic devices in SiC. The significant progress achieved in recent years now makes it of interest to publish a book specifically about process technology for SiC devices, to enable more researchers to investigate these devices.

This book is divided into seven chapters. The first chapter discusses the material properties of SiC, and specifically the advantages of SiC that started the interest in the first place. This chapter also includes some basic calculations on high-voltage blocking and on-resistance. Chapters 2–6 cover the basic process steps used in fabricating SiC devices. The chapters cover bulk and epitaxial growth of SiC (2), ion implantation and diffusion (3), wet and dry etching (4), thermally grown and deposited dielectrics (5), and Schottky and ohmic contacts (6). Lithography is not specifically covered, as this can be found in many standard books on silicon process technology [1]. The final chapter, Chapter 7, covers devices in SiC, divided into different categories. Here the combination of process steps from previous chapters is considered, along with the advantages of using SiC discussed in the first chapter. The first section of Chapter 7 on process integration may be read before Chapters 2–6. For details on device operation and semiconductor physics, the reader should refer to other books [2].

This book has been a cooperative effort, and as editor I have had the pleasure of working with several talented people. The credit for the good work is theirs, but the final responsibility for remaining errors is mine.

Carl-Mikael Zetterling
Stockholm, June 2002

REFERENCES

[1] J.D. Plummer, M.D. Deal, P.B. Griffin [*Silicon VLSI Technology* (Prentice Hall, 2000)]

[2] S.M. Sze [*Physics of Semiconductor Devices* 2nd edn (Wiley-Interscience, 1981)]

Chapter 1

Advantages of SiC

C.-M. Zetterling and M. Östling

1.1 CHAPTER SCOPE

This chapter will describe the basic properties of SiC, and show the advantages of using SiC over other semiconductor materials. First the crystallographic properties will be described, so that the anisotropy in the electrical properties can be understood. Some other important non-electrical properties are covered as well. The main part of the chapter covers the electrical properties relevant for electronic devices, and this section will reveal the advantages of SiC.

1.2 MATERIAL PROPERTIES

1.2.1 Introduction

Compared with other semiconductor materials, silicon carbide has the sometimes confusing property of occurring naturally in several different crystal structures, called polytypes. After a description of what polytypes are, and how they occur, some of the directly derived properties, such as lattice parameters and bandgap, will be discussed. Thermal expansion and hardness and other non-electric properties such as dielectric constant will be covered as well.

1.2.2 Tutorial

Crystals are ordered configurations of atoms occurring in all materials in the solid state. Most crystal structures can be described using a unit cell, which is repeated to describe the locations of all atoms in the crystal. For instance, silicon and carbon occur in the diamond unit cell. When a crystal is comprised of two atom types, like silicon (Si) and carbon (C) to make SiC, it becomes a little more complicated. Normally, every other atom in a diamond lattice is Si, and every other is C, so that each Si atom has exactly four C atom neighbours, and vice versa. The resulting crystal is called zinc-blende or sphalerite (FIGURE 1.1).

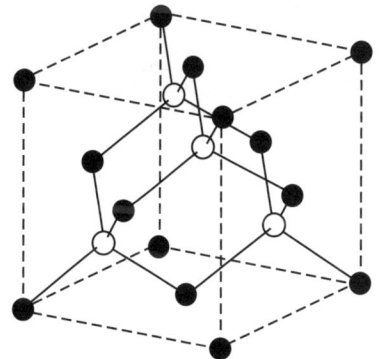

FIGURE 1.1 The zinc-blende crystal structure. White atoms are Si and black atoms are C [1].

1

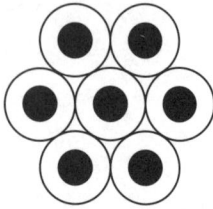

FIGURE 1.2   One double layer of hexagonally close packed atoms. White atoms are Si and black atoms are C.

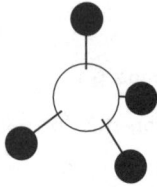

FIGURE 1.3   Si atom (white) surrounded by four C atoms (black) in a covalent bonding scheme.

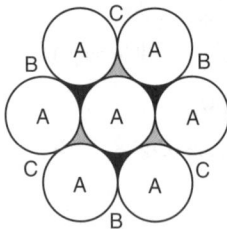

FIGURE 1.4   The position of the first layer of atoms is referred to as A. The next layer can either occupy positions B (black) or C (grey).

However, rather than describing a crystal starting from a cubic unit cell, it is also possible to start from a hexagonal close packing scheme. When perfect spheres are packed close to each other in a plane, they take positions in patterns that are hexagonal. Imagine a layer of Si atoms stacked this way, with a layer of smaller C atoms directly on top (FIGURE 1.2). The next double layer will then take positions where the atoms are centred between three atoms in the underlying layer. The reason for this is that Si and C prefer covalent bonds, which means that every atom has exactly four neighbours, and the atom bonds are in four directions (FIGURE 1.3).

When the third double layer of Si and C atoms is to be placed, it can be placed directly above the first layer. If this is repeated for every other layer, the resulting crystal structure is called wurtzite. However, the third layer may also be stacked in a different position from either the first or the second layer. To bring order to this, let us give the layer positions names, as in FIGURE 1.4. The wurtzite structure can then be described by the repeating sequence ABABAB etc., or simply AB. Since it takes two layers for this to repeat, and it is a hexagonal structure, this is commonly referred to as 2H SiC. If we instead stack the layers ABC etc., we have a cubic crystal when viewed along an axis diagonal to the hexagonally close packed plane, and it is referred to as 3C. Other possibilities are ABAC, called 4H, and ABCACB, called 6H. Each of these separate stacking orders is called a polytype. 3C SiC is sometimes referred to as beta-SiC, whereas all the other polytypes are referred to as alpha-SiC (FIGURE 1.5).

Note that each layer is a double layer of Si and C atoms, even though we describe the stacking using spheres. Since we can never stack layers of the same letter on top of each other (A on A, B on B etc.) one might be misled into believing that there are

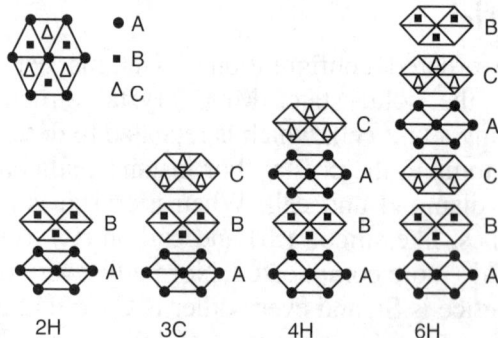

FIGURE 1.5   Some common crystal polytypes, 2H, 3C, 4H and 6H.

not so many possible polytypes. However, not only are several hundred stacking orders possible, they have also been identified in nature. For device purposes it is necessary that the wafer is a single crystal, and only a few polytypes are stable enough that large wafers have been made. Commercial wafers today can be bought as either 4H or 6H SiC, with diameters of 50, 75 or even 100 mm. Early experiments were often made on 3C SiC, which had been grown on Si wafers, but this material is not as good as the single crystalline wafers now available. For more details on the growth of bulk SiC material, see Chapter 2.

1.2.3 Applicability

For cubic crystals, three Miller indices, hkl, are used to describe directions and planes in the crystal. These are integers with the same ratio as the reciprocals of the intercepts with the x-, y- and z-axes, respectively (FIGURE 1.6) [1]. For instance, the (100) plane is one of the six surfaces of the cube, whereas the (111) plane is perpendicular to the volume diagonal. For hexagonal crystal structures, four principal axes are commonly used: a_1, a_2, a_3 and c. Only three are needed to unambiguously identify a plane or direction, since the sum of the reciprocal intercepts with a_1, a_2 and a_3 is zero. The three a-vectors (with 120° angles between each other) are all in the close-packed plane, also called the a-plane, whereas the c-axis is perpendicular to this plane (FIGURE 1.7).

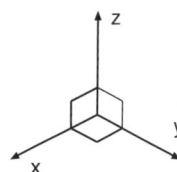

FIGURE 1.6   Principal axes for cubic crystals.

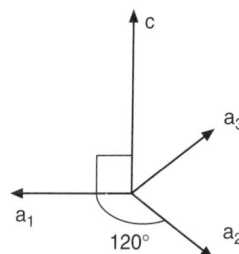

FIGURE 1.7   Principal axes for hexagonal crystals.

SiC has a polar crystal structure. Looking perpendicularly at the a-plane, we will either see C atoms directly on top of Si atoms, or vice versa. The former is called the silicon face orientation, the latter the carbon face. The silicon face is the face most commonly polished and used to manufacture devices on.

The spacing between the Si and C atoms determines the lattice parameters. Since the density is the same for all SiC polytypes, the atom spacing is the same. In the a-plane the lattice parameters are therefore the same, but along the c-axis the lattice parameter depends on the number of layers in the unit cell. For the cubic polytype the lattice parameter is different, since the unit cell is oriented differently, but within the crystal the atom spacing is still the same.

The lattice parameters are used to determine suitable substrates for epitaxial growth of SiC, or suitable films for epitaxial growth on top of SiC. Apart from the lattice parameters, the thermal expansion coefficients need to be closely matched, since epitaxial growth is done at high temperatures. If there is a large mismatch between these parameters for the two materials, a large strain will develop during cooling, which can crack the

TABLE 1.1   Mechanical properties of SiC and other semiconductors.

	(Unit)	Si	GaAs	3C-SiC	6H-SiC	4H-SiC	2H-GaN	Diamond
E_g	(eV)	1.12	1.43	2.4	3.0	3.2	3.4	5.5
Lattice a	(Å)	5.43	5.65	4.36	3.08	3.08	3.189	3.567
Lattice c	(Å)	n.a.	n.a.	n.a.	15.12	10.08	5.185	n.a.
Bond length	(Å)	2.35	2.45	1.89	1.89	1.89	1.95	1.54
T.E.C.	(10^{-6}/K)	2.6	5.73	3.0	4.5		5.6	0.8
Density	(g/cm^3)	2.3	5.3	3.2	3.2	3.2	6.1	3.5
Th. cond. λ	(W/cm K)	1.5	0.5	5.0	5.0	5.0	1.3	20.0
Melting point	(°C)	1420	1240	2830	2830	2830	2500	4000
Mohs hardness				9.0	9.0	9.0		10.0

T.E.C. = thermal expansion coefficient

film or cause relaxation (broken bonds) at the interface unless the film is kept thin. For more details on epitaxial growth, see Chapter 2. The mechanical properties of SiC are summarised in TABLE 1.1.

Since the hexagonal polytypes are made up of stacked double layers, it is not surprising to find that several material properties are different along the c-axis or perpendicular to the c-axis (in the a-plane). This is called anisotropy, and the degree of anisotropy is measured by the quotient of a parameter value along and perpendicular to the c-axis. An anisotropy of 1 is the same as isotropy, and it is only 3C SiC that is isotropic. Several of the electrical parameters are anisotropic: see Section 1.3.3.

The density of SiC is not very high, but considering that both Si and C are light atoms, it turns out that the atomic density is large. The number of Si atoms per cubic cm is the same in SiC as in Si, but SiC also has the same number of C atoms squeezed into the lattice. With such short bond lengths between the atoms, it is not surprising that the bond strength and hence the hardness is very high. Only diamond is harder, which makes SiC wafers difficult to polish.

The short bond length also means that SiC has a wide energy bandgap. The energy bandgap is the smallest energy needed to excite an electron from the valence band to the conduction band. This energy has a temperature and doping dependence, which will be discussed in Section 1.3.3. The bandgap can be classified as direct or indirect, and direct bandgap semiconductors have a higher efficiency in emitting photons. SiC is an indirect bandgap semiconductor. Surprisingly, the energy bandgap is quite different for the different polytypes of SiC, varying from 2.2 to 3.4 eV. However, the optical properties do not vary much between the polytypes: see TABLE 1.2.

1.2.4 Overcoming problems

As mentioned before, the lattice match and thermal expansion coefficients are important when epitaxial growth is performed with or on SiC: see the details in Chapter 2. This can cause large stress in epitaxial films, which leads to cracking or warpage of the wafer. The wafer can also become warped after too rapid heating or cooling, or if the bulk SiC has built-in warpage when it is cut into wafers. During processing there can occur extra deposition or growth along the edges of the wafer, so called epi-crown. All of these problems can cause difficulties with the lithography, if the surface cannot be kept parallel to the exposure plane.

The hardness of the material makes polishing of SiC wafers difficult. Although the surface can be partially removed using hydrogen etching (see Chapter 2 or 4) or sacrificial oxidation (see Chapter 5), coarse scratches are difficult to remove completely. SiC wafers will also scratch any glass parts in the process machinery used; for instance some mask aligners press the wafer into contact with either the mask directly, or a glass plate, and scratches will quickly develop in these.

Defects can be present in the material from the growth, for instance micropipes and screw dislocations (see discussion in Chapter 2), or can be created during processing (see Chapter 3 on ion implantation) or even during operation of the devices (see 7.3.4). These are the main concerns for any device manufacturing in SiC.

1.3 ELECTRICAL PROPERTIES

1.3.1 Introduction

There are several good reasons to choose SiC as a semiconductor device material, but the relevant electrical property depends on the device application intended. Some important device parameters in terms of the electrical properties of the semiconductors will be discussed. The different polytypes of SiC have an energy bandgap at room temperature varying from 2.2 eV for 3C SiC to 3.4 eV for 4H SiC. The wide bandgap is the key to almost all advantages of using SiC in devices. All properties are summarised in TABLE 1.2. The detailed device structures will be discussed in Chapter 7.

TABLE 1.2   Electrical properties of SiC and other semiconductors.

	(Unit)	Si	GaAs	3C-SiC	6H-SiC	4H-SiC	2H-GaN	Diamond
E_g	(eV)	1.12	1.43	2.4	3.0	3.2	3.4	5.6
E_c	(MV/cm)	0.25	0.3	2.0	2.5	2.2	3.0	5.0
v_{sat}	(10^7 cm/s)	1.0	1.0	2.5	2.0	2.0	2.5	2.7
$\mu_{n,\perp c}$	(cm^2/V s)	1350	8500	1000	500	950	400	2200
$\mu_{n,\parallel c}$	(cm^2/V s)	n.a.	n.a.	n.a.	100	1150	n.a.	n.a.
μ_p	(cm^2/V s)	480	400	40	80	120	30	1600
ε_r		11.9	13.0	9.7	10.0	10.0	9.5	5.0
Th. cond. λ	(W/cm K)	1.5	0.5	5.0	5.0	5.0	1.3	20.0

1.3.2   Tutorial

1.3.2.1   Wide bandgap – high temperature advantage

The energy bandgap is the minimum energy needed to excite an electron from the valence band to the conduction band. Thermal energy through lattice vibrations can create electron-hole pairs in the semiconductor even at room temperature. If the temperature is high enough, more electron-hole pairs can be created than the number of free carriers from the impurity doping. When this happens the material becomes intrinsic and the devices fail, as there is no longer any p-n junction to block the voltages. For high-temperature devices, higher doping may be used to raise the threshold, the intrinsic temperature, where thermal generation is too high. SiC with its wide bandgap has an intrinsic temperature around 1000 °C, depending on polytype and doping. At these temperatures the devices are more likely to fail because of contact failure. However, although there are many high-temperature applications, there has not been a breakthrough in this field. There are problems in making long-term reliable devices and packages for them at extended temperatures, and the electronic circuits need to be designed to take varying threshold voltages into account.

1.3.2.2   Wide bandgap – optical advantage

The wide bandgap means that SiC has optical detection capability in the UV range. At the same time the sensitivity is small for light with lower energy, hence longer wavelengths. However, since SiC has an indirect bandgap (the minimum in the conduction band does not coincide with the maximum in the valence band), optical emission is inefficient. Although LEDs with blue light were made initially in SiC, this market is dominated nowadays by GaN, which has a direct bandgap.

1.3.2.3   Wide bandgap – high electric field advantage

Because of the wide bandgap, the impact ionisation energy is high in SiC. This means that the electric field can become very high without avalanche multiplication of ionised carriers. In Si crystals an electric field of 0.25 MeV/cm is sufficient to accelerate carriers so that they will generate more carriers through impact ionisation. For SiC the electric field can be ten times higher before this happens. The exact electric field needed depends slightly on the doping concentration, and an exact calculation needs to take this into account. For a device design a safety margin is used, so perhaps 70% of the maximum electric field should be used in the calculations.

The highest electrical field occurs in the blocking p-n junction of the device, and it is this field that determines the breakdown voltage. Normally the blocking junction is one-sided, meaning that one side has much higher doping. The depletion region will mainly spread into the low-doped side. If we assume a uniform doping on the low-doped side, the electrical field will take the shape seen in FIGURE 1.8. The blocking voltage, also called the breakdown voltage V_B, is the product of the critical field E_C and the depletion width W, divided by two, EQN (1):

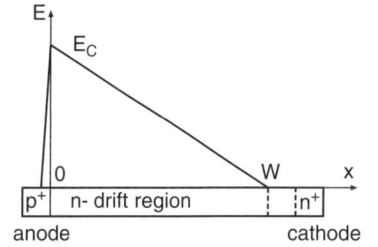

FIGURE 1.8   Electric field in an abrupt one-sided p-n junction prior to breakdown.

$$V_B = \frac{E_C W}{2} \tag{1}$$

This means that we can make devices with ten times higher breakdown voltage than Si using the same depletion width.

The low-doped region should be thick enough to accommodate the depletion region, and this is called a non-punch-through (NPT) design. However, since the doping is low in this region, this will cause a large voltage drop, so the width should be minimised. In a punch-through (PT) design, the depletion width is slightly larger than the low-doped region. From EQN (1) we can see the advantage of SiC: for the same breakdown voltage design we can have a ten times thinner depletion region. Next we can calculate the doping N_D for this low-doped region from the following relation, EQN (2):

$$W \approx \sqrt{\frac{2\varepsilon V_B}{qN_D}} \quad \Rightarrow \quad N_D = \frac{2\varepsilon V_B}{qW^2} = \frac{\varepsilon E_C^2}{2qV_B} \tag{2}$$

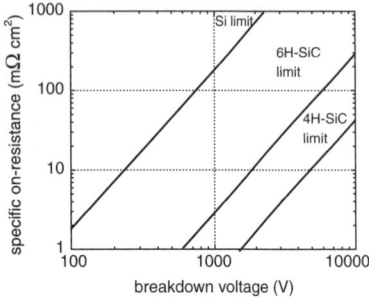

FIGURE 1.9 On-resistance for a blocking region designed for a certain breakdown voltage in Si, 6H SiC and 4H SiC.

Consequently the doping in the low-doped region for SiC can be 100 times higher for the same breakdown voltage. Finally we can calculate the on-resistance R_{on} for this low-doped region:

$$R_{on} = \frac{W}{q\mu_n N_D} = \frac{4V_B^2}{\varepsilon\mu_n E_C^3} \tag{3}$$

With a ten times thinner depletion region and 100 times higher doping, the on-resistance is approaching 1000 times lower values. However, since the electron mobility μ_n and dielectric constant ε are lower for SiC, the advantage is between 200 and 400, depending on polytype. This is the main advantage of SiC high-voltage devices: the on-resistance is much lower for the same breakdown voltage (FIGURE 1.9).

1.3.2.4 Wide bandgap – high-frequency advantage

As we saw in the previous section, the devices can be made smaller for the same breakdown voltage. This means of course that the device is faster, since the signal has a shorter distance to travel. The relative dielectric constant is also lower for SiC than for most other semiconductors, which means that parasitic capacitances will be smaller, since the capacitance is directly proportional to the dielectric constant. Semi-insulating wafers are available in SiC, although they are still very expensive. The saturated electron velocity is also high in SiC, twice that of Si and GaAs. A disadvantage is that the electron mobility is smaller than in the other semiconductors.

1.3.2.5 High thermal conductivity – high power advantage?

It is often stated that the high thermal conductivity λ of SiC allows higher power densities. Being three times higher, this means that three times higher thermal flow θ can be accommodated with the same temperature increase ΔT at the junction, according to EQN (4) (t is the device thickness):

$$\Delta T = \theta\frac{t}{\lambda} \tag{4}$$

However, considering that the devices are made ten times smaller, if the advantage of the higher critical field is exploited, we should expect to have ten times higher power flow θ. This means that we should also expect to see higher junction temperatures for SiC devices, even though the thermal conductivity is higher. We should also remember that often we are limited by the

8

thermal conductivity of the solder mounting of the chip to the carrier.

1.3.3 Applicability

Physical device simulation is used frequently to predict characteristics of electronic devices, either in the design phase before devices are manufactured, or in an exploratory phase to investigate the advantages of possible devices. Although it is important to have as accurate models as possible, sometimes all model parameters are not available for the case in question. By simplifying the simulation, results can be reached faster. Some of the more important simulation problems are described in the following paragraphs. Simulation is important, both to design devices and to understand the operation of manufactured devices.

None of the properties in TABLE 1.2 are constants. Over a large temperature interval they all show variations, but for a simulation at a fixed temperature this is normally not a problem. Heavy doping will alter properties as well, since the crystal will be less like SiC when dopant atoms replace a large portion of the atoms. Models do not exist for all cases that need to be simulated, and often models have to be used that were originally developed for other semiconductor materials. It is worth looking into the models used by the simulation programs closely: they are normally documented in the user manuals. Another general reference on SiC modelling is [2]. The difference in bandgap energy between the polytypes of SiC has been found to be almost entirely in the conduction band: the valence band coincides.

One difficulty with SiC is the low intrinsic carrier concentration n_i at room temperature, much less than one per cubic cm (EQN (5)):

$$n_i = \sqrt{N_C N_V} \left(\frac{T}{300}\right)^{3/2} \exp\left(-\frac{E_g}{2kT}\right) \qquad (5)$$

This is caused by the wide bandgap, and has nothing to do with the simulator. The minority carrier concentration at equilibrium is calculated from the law of mass action, EQN (6):

$$np = n_i^2 \qquad (6)$$

When the simulator has to cope with majority carrier concentrations on the order of $10^{18}\,\mathrm{cm}^{-3}$ and minority carrier concentrations less than $10^{-18}\,\mathrm{cm}^{-3}$, the numerical accuracy is not enough, and cancellation problems are common. One way around this is to start the simulations with the temperature set to around 600 K,

9

when the influence of small numbers of minority carriers is present, for instance in leakage current calculations or avalanche-breakdown calculations. Once the avalanche condition has been reached, the temperature can be lowered.

A similar problem occurs with deep traps. At room temperature the filling and emptying time is slow because of the large activation energies. However, this is also the case for measurements, so elevated temperatures can be used in simulation and measurement. High-frequency simulations can also be difficult to perform, since convergence problems often occur. If minority carriers are expected to influence the result, raising the temperature might help in this case as well.

Earlier in this chapter it was mentioned that due to the anisotropic nature of the SiC crystal structure, anisotropic electric properties should be expected. This means that electrical characteristics will be different depending on the orientation of the device with respect to the crystal. Mobility is one prime example. For 6H SiC the electron mobility is five times lower parallel to the c-axis, compared to perpendicular to the c-axis. The most commonly used orientation has the wafer surface perpendicular to the c-axis, which means that the current transport is better in a lateral device compared with a vertical device. Unfortunately, most power devices depend on vertical current transport, since it is easier to manufacture a blocking layer parallel to the surface (see FIGURE 1.10). This was one main reason for many research groups switching to the 4H SiC polytype, which has higher mobility parallel to the c-axis, and much smaller mobility anisotropy (20%).

The difficulty is that not all numerical device simulators support anisotropy for the electrical or material properties, since they have originally been made for isotropic materials like Si and GaAs. For anisotropy of the order of 20%, usually an approximation of SiC as an isotropic material is sufficient, especially if worst-case parameters are used in all directions. The impact ionisation coefficients also have some anisotropy, and this will, of course, be important when calculating the breakdown voltage [3]. However, a safety margin is always used for the device design with respect to breakdown voltage. Since this anisotropy is not as large as the 6H electron mobility anisotropy, a safe route is to use the worst-case parameters in all crystal orientations, and do the calculations approximating SiC as an isotropic material.

Another difficulty is the channel inversion mobility in MOSFETs: see Chapter 5. Initially the mobility was believed to be extremely low, a few per cent of the bulk mobility in SiC. For comparison, the inversion mobility in Si is slightly less than 50% of the bulk mobility. However, it has been found that due to

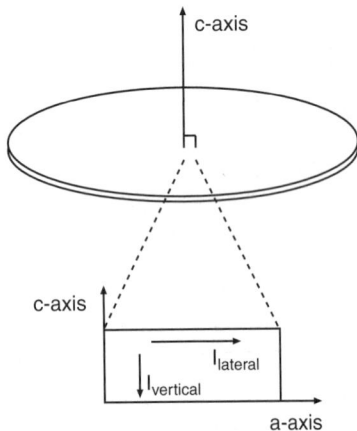

FIGURE 1.10 Wafer with surface perpendicular to the c-axis: current transport laterally and vertically defined.

large numbers of traps in the channel, 80% of the channel charge is fixed. Therefore the mobility extraction has underestimated the true mobility by a factor of five or more, compared with trap-free devices.

The final important matter to take into account when simulating SiC is the incomplete ionisation of dopants at room temperature, which is described in detail in Chapter 3. All dopants in SiC are much deeper in the bandgap compared with Si, and this means that at room temperature not all dopants are ionised, and hence the free carrier concentration is not equal to the dopant concentration. This is especially the case for p-type dopants. The problem is also larger for higher amounts of doping, at least until the material is degeneratively doped (above $10^{19}\,cm^{-3}$) in which case the bands are broadened and approximately all dopants become ionised. Incomplete ionisation can lead to some unexpected effects; for instance the current gain of bipolar transistors will drop when the base doping becomes ionised at higher temperatures.

1.3.4 Overcoming problems

The problems of dopant ionisation and activation of dopants are discussed in greater detail in Chapter 3. Working with physical device simulation is difficult, even when new materials such as SiC are not included. Apart from the general advice of simulating with initially elevated temperature for breakdown, the only way to learn is to do many simulations and be critical of the results and the models used. The results can never be better than the input to the device simulator.

1.4 CONCLUSIONS

This chapter has discussed some properties of SiC and especially those that differ significantly from those of other semiconductors. Due to the anisotropic crystal structure of SiC, several properties display anisotropy as well. The wide bandgap is perhaps the most important property, since it allows significantly higher critical fields in the devices. Thus smaller devices can be made, which have lower on-resistance and are faster. This is the main reason for all the work that has been done on SiC growth, processing and device design. It is necessary to use a device simulator for design of SiC devices, and several models have to be adapted for SiC. The low intrinsic carrier concentration, which is the reason for low leakage currents in SiC, also makes numerical simulation difficult due to the variation over many orders of magnitude in the carrier concentration.

REFERENCES

[1] C. Kittel [*Introduction to Solid State Physics* 7th ed. (Wiley, 1995)]

[2] M. Bakowski, U. Gustafsson, U. Lindefelt [*Phys. Status Solidi A (Germany)* vol.162 (1997) p.421–40]

[3] O. Tornblad, M. Östling, U. Lindefelt, B. Breitholtz [*Semicond. Sci. Technol. (UK)* vol.14 (1999) p.125–9]

Chapter 2

Bulk and epitaxial growth of SiC

N. Nordell

2.1   CHAPTER SCOPE

Bulk crystal growth is the technique for fabrication of single crystalline substrates, making the base for further device processing. This chapter describes the seeded sublimation technique used today for high-quality crystals. The focus is on the main growth parameters that are important for doping control and defect reduction. Also, liquid phase epitaxy and the recently presented high-temperature chemical vapour deposition technique for bulk crystals are discussed.

In order to fabricate doped device structures, which demand extremely high crystal quality with low defect density, vapour phase epitaxy is performed on the substrates grown by bulk techniques. Vapour phase epitaxy is also treated in detail, with a description of the most common reactor geometries, precursor gases and growth parameters. Methods to avoid defects and to improve the doping profiles are also presented.

2.2   BULK GROWTH

2.2.1   Introduction

A major breakthrough for SiC as a material for semiconductor devices came when substrates could be produced with a reproducible process, which could be scaled up to large substrate sizes. Today, substrates up to 50 mm in diameter are commercially available from several suppliers, and 100 mm substrates have been reported. The initial problems with inclusions of foreign polytypes are basically solved, and the defect concentration is steadily decreasing. The doping concentration can be controlled over a wide range. In addition to n- and p-type doping, semi-insulating substrates are offered.

The seeded sublimation technique, where the crystal is grown from a solid source, has proven successful and is today preferred

13

for substrate fabrication. However, this method has to be improved with respect to micropipe defects, background doping and mechanical stress in the wafers. None of these issues are inherent in the technique, and they should all be eliminated by further process optimizations.

For most semiconductor materials, bulk crystals are grown from a liquid source. This method is also possible to use for SiC, and has some advantages, mainly being more forgiving to defects. However, this method is still only used for small-scale experiments and a considerable development effort is needed before it could be used for substrate production.

Also, a gas-phase growth technique, high-temperature chemical vapour deposition, has been reported. It offers the possibility to use high-purity precursors and independently controlled process parameters. This process might have the potential to be competitive with the sublimation technique in the future.

2.2.2 Tutorial

2.2.2.1 *Sublimation growth techniques*

The first commercial process to synthesize SiC was established by Acheson in 1891. The Acheson process has been refined and is still used for its original purpose: to produce SiC for abrasive applications. FIGURE 2.1 shows a schematic of an Acheson furnace, which could be between 6 and 50 m long and loaded with up to 3000 tons of the basic raw materials, petroleum coke and silica sand, surrounding a graphite core. The core acts as heating element when energised by electricity at high current and low voltage, but is also partly consumed during the process. The furnace is heated to 1700–2500 °C for 7–10 days, allowing SiC to form, with CO as the main by-product. In the hot centre of the

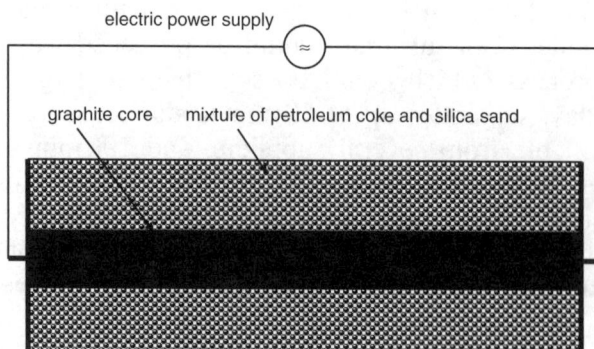

FIGURE 2.1   A schematic of an Acheson furnace.

furnace (>2100 °C) α-SiC is produced, usually in the form of crystalline platelets with hexagonal shape, while amorphous and β-SiC occur at the outer edges. The α-SiC is preferred as abrasives when it is ground to a powder. The quality depends on the purity, which mainly relates to the purity of the raw materials [1].

With the goal to grow SiC crystals with high purity and good crystallinity for semiconductor purposes, Lely developed a small-scale sublimation technique in 1955. The original Lely process employs a cylindrical graphite crucible of a hundred cubic centimetres in volume, with the inside walls covered by SiC lumps (FIGURE 2.2), e.g. produced by the Acheson process. When the crucible is heated to the process temperature of about 2550 °C, volatile Si-C species evaporate and SiC platelet crystals, similar to those formed in the Acheson process, are grown attached to the lumps at the walls inside the crucible, while dense layers of SiC are formed at colder surfaces [1].

The hexagonal platelets formed in the Acheson and Lely processes are of dimensions up to $10 \times 10 \times 3 \, mm^3$. They consist mainly of 6H, 4H or 15R SiC, and it is not uncommon to have mixed polytypes within the same platelet. The main surfaces of the platelets are the Si- and C-terminated crystallographic c-planes (FIGURE 2.3). The colour depends on polytype and doping (TABLE 2.1).

The difficulty in controlling the polytypes obtained in the Lely process made Tairov and Tsvetkov develop the seeded sublimation growth or the modified Lely process [2]. This process is performed in a crucible similar to the one used by Lely, with a SiC source, and in addition a single crystal seed, e.g. a Lely or Acheson crystal platelet, is fixed inside the crucible. During the process the temperature of the source is about 100 °C higher than that of the seed (which is 1800–2400 °C). This makes the volatile Si-C species leave the source and condense on the colder seed, where the single crystal SiC is growing epitaxially.

The seeded sublimation growth method has been refined and is today used for fabrication of most SiC substrates. Through consecutive improvements of the crystals, e.g. by a careful

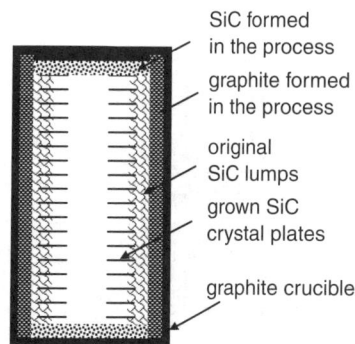

FIGURE 2.2   A schematic of a crucible for Lely growth.

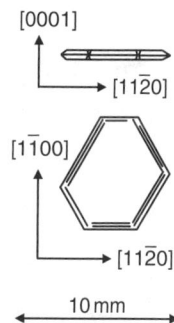

FIGURE 2.3   Hexagonal SiC platelet, with the major crystallographic directions shown.

TABLE 2.1   Colours of some SiC polytypes, depending on doping.

	4H	6H	15R	3C
n >1E18	brown	green	yellow	yellow
p >5E18	blue	blue	blue	green
Low doped	transparent			
Heavily doped	black			

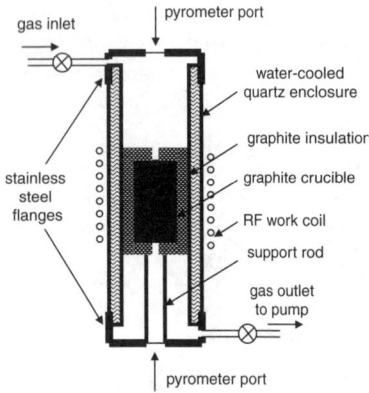

FIGURE 2.4   A schematic of a furnace for seeded sublimation.

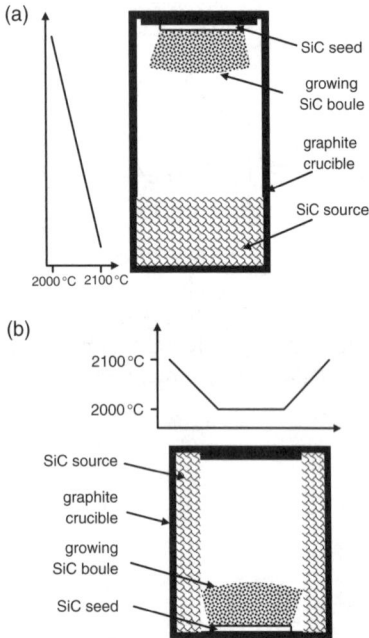

FIGURE 2.5   Cross section of a crucible for sublimation, with the typical temperature profile shown:
(a) lid mounting of the seed;
(b) bottom mounting of the seed.

choice of seed crystals in order to reduce the defect concentration, the substrate quality of today has been developed.

2.2.2.2   *The seeded sublimation growth process*

The typical equipment used for seeded sublimation growth consists of a water-cooled quartz reactor enclosure, surrounding the graphite crucible, in which the process takes place (FIGURE 2.4). The crucible is heated by radio frequency (RF) induction, and is thermally insulated and shielded by a graphite felt or porous graphite. The RF frequency is chosen in such a way that it does not heat the graphite insulation. A pump and a gas source (usually argon) are connected to the reactor. In order to measure and control the process temperature and the temperature gradient of the crucible, pyrometer ports can be opened through the insulation at the top and bottom of the crucible.

The crucible has a diameter somewhat larger than the diameter of the wafer to be grown and is partly filled with SiC powder with particle sizes in the order of 20–200 μm, or with sintered polycrystalline SiC. Typically, a distance of between 1 and 20 mm is allowed between the SiC source and the seed crystal. The seed could be fixed either to the lid of the crucible by clamping or gluing, e.g. by melting high-purity sugar (FIGURE 2.5(a)), or to the bottom of the crucible, in which case the SiC source is surrounding the seed (FIGURE 2.5(b)).

At the process temperature of typically 1800–2400 °C, the volatile species of Si_2C, SiC_2 and Si evaporate from the source, and condense on the seed substrate. The substrate forms a template for crystal growth, and a single crystal grows on the substrate according to the template. The main driving forces for the growth are the source temperature, which determines the rate of evaporation from the source, and the temperature difference between the source and seed, which determines the diffusion transport rate of the species from the source to the seed. An increased growth rate will be the result, both when the source temperature and when the temperature difference is increased. On the other hand, an increased seed temperature will enhance the re-evaporation from the seed and could lead to a reduced growth rate.

It has been shown that the growth temperature influences the relation between SiC_2 and Si_2C in the gas phase, with SiC_2 dominating at higher, and Si_2C at lower temperatures. At all temperatures, the gas phase is Si-rich, due to depletion of Si from the source, where C accumulates. Also, graphite from the crucible itself evaporates and participates in the reaction. This results in deposits of C-rich SiC and graphite on the hottest parts of the crucible, and C inclusions can be incorporated in the growing SiC. A more accurate control of the growth parameters is obtained in a closed crucible, where the Si-rich gas phase does not leave the crucible.

Under normal growth conditions a Si:C ratio significantly above unity is desirable in order to avoid graphitization of the growing surface. In order to obtain this, Si can be added to the SiC source. Simultaneously, this increases the probability for the occurrence of a liquid Si phase, which does not harm the growth unless Si droplets are formed at the growing crystal surface [3].

The temperature optimization has to be made with care, as it influences not only the growth rate, but also the thermodynamic growth conditions. The process temperature has to be chosen in a way to avoid graphitization or formation of silicon droplets on the growing surface. In addition, the silicon content of the source has to be adjusted to the growth temperature. It is especially important to minimise the temperature distribution over the source or seed, in order to avoid local variations, which could enhance unwanted inclusions. The detailed adjustment of the temperature field has to be made for each specific geometry.

The sublimation process is performed in vacuum or in an ambient of an inert gas, e.g. argon. Both nitrogen and hydrogen are strongly reactive at optimum growth temperatures, and should be avoided.

The process pressure is a main parameter to control the sublimation growth rate. As the transport of species from the source to the seed is diffusion limited, decreasing the pressure will enhance the diffusion of growth species through the gas phase. The resulting growth rate is practically inversely proportional to the pressure, down to around 10^{-1} mbar, where the convective transport of species determines the growth rate, FIGURE 2.6 [4]. Most groups work at pressures below 100 mbar and down to 10^{-1} mbar. The minimum pressure that is possible to reach is limited by the vapour pressure of the reacting species at the growth temperature, and the intention to keep a Si-rich ambient.

In the diffusion-limited growth regime the growth rate also shows a linear dependence on the inverse distance between the source and seed (FIGURE 2.7) [5].

The typical growth sequence starts with de-gassing the crucible, with the source and seed in place. This process is performed in an inert gas ambient at atmospheric pressure and at a temperature close to or even above the growth temperature. Through this procedure, any gases or volatile impurities adsorbed in the material will leave the crucible. In order to prevent deposition on the substrate, the temperature gradient in the crucible is reversed as compared to the growth conditions, which will also slightly etch

Dominating reactions during sublimation growth [3]:

Solid SiC forms from gas components:
$SiC_2(g) + Si_2C(g)$
$\leftrightarrow 3\ SiC(s)$

SiC_2 transforms into Si_2C in gas phase:
$SiC_2(g) + 3\ Si(g)$
$\leftrightarrow 2\ Si_2C(g)$

Solid C forms from gas components:
$Si_2C(g) \leftrightarrow C(s) + 2\ Si(g)$

Liquid Si forms from gas components:
$Si(g) \leftrightarrow Si(l)$

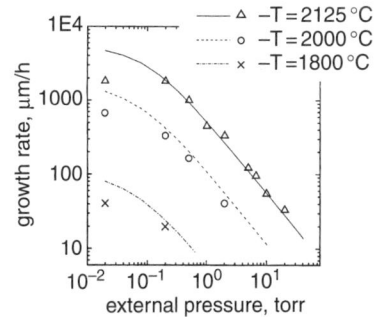

FIGURE 2.6 Growth rate vs. gas ambient pressure for different temperatures. From [4].

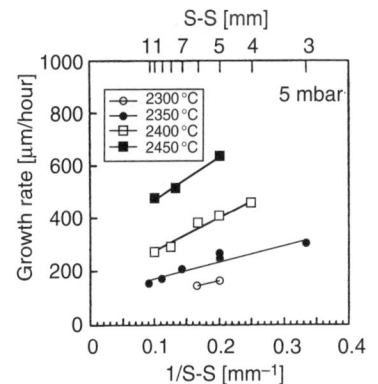

FIGURE 2.7 Growth rate vs. source-seed distance for different temperatures. From [5].

the substrate surface, forming a clean surface for subsequent growth.

After de-gassing, the growth parameters are adjusted to those intended for growth by adjusting the temperature and the temperature gradient, and finally the pressure is decreased in order to start the deposition process. The maximum growth rates for high-quality crystals can reach up to 0.5–1.0 mm/h, and the growth continues for up to 40 h, producing a boule of a diameter limited by the crucible and of 20–40 mm length.

2.2.2.3 Dopant incorporation

Nitrogen and aluminium are the dopants of preference for n- and p-type crystals, respectively. In order to obtain high resistivity or semi-insulating material vanadium is added.

Nitrogen doping is obtained by adding nitrogen gas during growth. The nitrogen incorporation is proportional to the square root of the vapour pressure of the N_2 gas in the crucible, and is practically independent of the growth rate [5]. This implies that the incorporation rate is determined by the equilibrium between nitrogen in the gas phase and nitrogen adsorbed on the surface of the growing crystal. It is also found that the N incorporation is 2–3 times higher for growth on the C face than on the Si face at 2300 °C, but this difference decreases at higher temperatures. The nitrogen doping can be controlled from 10^{17} cm^{-3} up to about 10^{21} cm^{-3}.

Aluminium doping is obtained by adding Al to the SiC source, or by using Al-doped source material. The Al incorporation is proportional to the Al vapour pressure in the gas, up to a saturation level at about 10^{21} cm^{-3} [6]. Contrary to N doping, the Al incorporation is higher for growth on the Si face than on the C face, at about 5 to 10 times at 2300 °C; this difference also decreases at higher temperatures (FIGURE 2.8).

B and Al impurities, which are both acceptors, are usually present in the source or crucible material in such a quantity that the p-type background doping or compensation level of n-doped material is in the 10^{17} cm^{-3} range. Vanadium forms deep levels in the bandgap, acting as traps for the holes. Hence, intentional addition of V has been shown to be a successful way to produce semi-insulating material, with resistivities above 10^8 Ω cm at 200 °C [7].

Doping incorporation is also sensitive to small variations of the crystallographic orientation, which makes it difficult to obtain uniformly doped bulk crystals, as the crystallization surface is usually non-planar, and a set of crystal planes is simultaneously exposed.

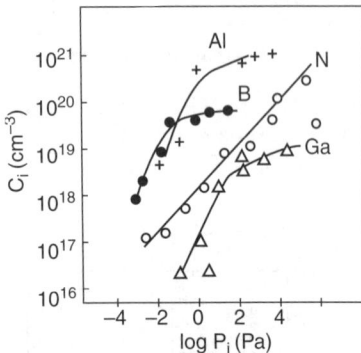

FIGURE 2.8 Doping vs. partial pressure of dopants. Al, B and Ga act as p-type dopants and N as n-type dopant. From [6].

2.2.2.4 *Polytype control and defect reduction*

The low stacking fault energy, and hence the high probability of forming lattice defects and inclusions of different polytypes, has been a major obstacle in bulk growth of SiC. Only with optimized growth parameters, will a superior crystal quality be obtained.

The main parameter determining which polytype is obtained seems to be the face of the seed crystal. This is true over a wide temperature range and regardless of polytype of the seed. Hence, it is found that growth on the (0001) Si face will give a 6H crystal, while 4H will grow on the (0001) C face. This could be explained by the large difference in surface energy between the faces [8]. Other important parameters are the growth temperature and pressure, where 6H is preferably grown at high pressure and at high temperature, while 4H is obtained at lower temperatures and pressures [9]. In addition, it has been observed that the intentional or unintentional incorporation of impurities can affect the polytype formation. Hence, nitrogen will favour more cubic polytypes, such as 3C and 15 R, also making nitrogen doping of 4H more difficult than 6H. On the other hand, rare-earth metals, germanium, tin, lead, and to some extent aluminium, favour polytypes with high hexagonality, i.e. primarily 4H [6]. Polytypic homogeneous samples are grown if supersaturation is not fluctuating at the growing surface and at a low growth rate.

The major obstacle in commercialisation of SiC devices is the high density of micropipe defects. Micropipes are hollow-core defects, with a typical diameter of 1–10 μm. During recent years the micropipe density has been reduced from more than 1000 cm^{-2} to less than 1 cm^{-2} in the best samples.

There are two major theories for explaining the mechanism of formation of micropipes. One attributes the defect formation to the incorporation of an inclusion, e.g. a graphite particle or a silicon droplet, during growth. The inclusion is believed to hinder the surface mobility of growth species, and hence gives rise to the hollow core. The other model explains the micropipe as being the centre of a screw dislocation with a large Burgers vector, where the diameter of the micropipe is related to the magnitude of the Burgers vector (FIGURE 2.9). These theories are based on observations; micropipes are found to emanate from inclusions in the crystal, and micropipes are also found to be in the centre of screw dislocations. It has been shown that these theories are compatible and that inclusions could cause screw dislocations (and micropipes), which propagate through the growing crystal [10]. The appropriate way to reduce

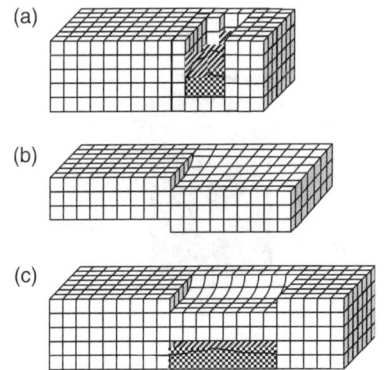

FIGURE 2.9   Possible mechanisms for micropipe formation: (a) inclusion giving rise to a hollow core; (b) screw dislocation (with the Burgers vector equal to one unit cell); (c) inclusion giving rise to two screw dislocations of opposite sign.

19

micropipes in the growing crystal is through accurate control of the growth parameters, keeping the conditions for nucleation stable at the growing surface, and hence avoiding deposition of liquid or solid phases, which would form nucleation sites for dislocations and micropipes.

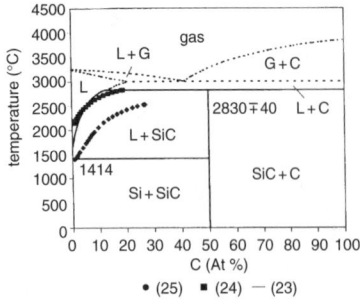

FIGURE 2.10   Phase diagram of the binary C-Si system. From [11].

2.2.2.5   Liquid phase epitaxy

For most semiconductor materials, bulk crystals are pulled from a melt. For SiC this method shows some extraordinary difficulties, as SiC only melts at pressures above 100 000 bar and at temperatures above 3200 °C. On the other hand, the Si-C phase diagram in FIGURE 2.10 shows that up to 15% of C could be dissolved in Si at 2800 °C [11]. By adding rare-earth or transition metals, such as Pr, Tb or Sc to the melt, the solubility of C can be increased up to 50%. In order to benefit from the high solubility, it is important to also consider additional aspects of the solvent, e.g. the incorporation of the solvent metal into the SiC crystal, the wetting properties of the crystal by the solvent, and that SiC should be the only stable phase in the system. From this perspective, the preferred solvent for growth of electronic material is pure Si. A drawback of using Si as the solvent is the high Si vapour pressure of up to 0.25 bar at 2300 °C. By applying a high pressure (up to 200 bar) of an inert gas, e.g. Ar, above the surface of the melt, the evaporation of Si can be reduced. Another issue has been the stability of the crucible material, as liquid Si is extremely reactive. By using graphite, the crucible material is allowed in the system, and the crucible can act as a carbon source for the growth. However, the incorporation of impurities from the graphite into the grown SiC will increase the background doping.

An experimental set-up for growth by liquid phase epitaxy (LPE) is shown in FIGURE 2.11. The furnace is resistivity heated and thermally insulated by graphite. Rotation of the melt is applied to ensure radial homogeneity of growth. The driving force for crystal growth is a slow cooling, leading to a continuously reduced solubility of C in Si, and a following nucleation of a solid SiC phase on the seed crystal [11]. The growth rates obtained are still below 0.5 μm/h, but it is probable that they will increase when the method has matured. It is possible to dope the material with Al for p-type by adding Al to the melt, or N for n-type by applying an atmosphere of nitrogen gas.

A major advantage with liquid phase epitaxy seems to be the polytype stability and the low density of micropipes formed during growth. It has even been shown to be possible to eliminate existing micropipes in substrates by overgrowing them.

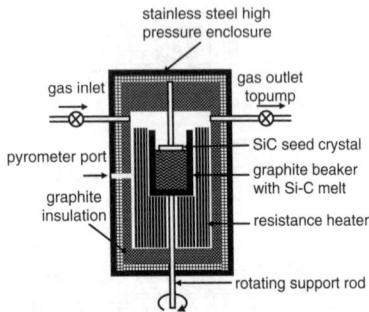

FIGURE 2.11   A schematic of a furnace for liquid phase epitaxy.

2.2.2.6 *High-temperature chemical vapour deposition*

A competitive alternative to the seeded sublimation growth and liquid phase epitaxy is high-temperature chemical vapour deposition (HTCVD). Growth takes place in a vertical reactor made of graphite (FIGURE 2.12), in shape similar to the crucible used for seeded sublimation. The seed is placed in the top, and instead of a solid source, the growth species consist of a stagnant supply of Si and C containing precursors, e.g. SiH_4, and a hydrocarbon, e.g. ethylene, C_2H_4, or propane, C_3H_8, diluted in a carrier gas [12]. By letting the precursors pass into a cracking zone, preferably heated by the walls of the reactor enclosure, reactive radicals are formed before reaching the seed. The temperature in the cracking zone should be slightly higher than that of the seed, in order to maintain a supersaturation in the gas phase at the surface of the seed. The heated reactor walls also facilitate re-evaporation, to avoid large SiC wall deposits. However, it is preferred to keep a thin, high-purity SiC coating of the reactor walls, in order to avoid the graphite evaporating, which would strongly increase the C : Si ratio in the gas phase, and also release the impurities bound to the graphite. The carrier gas should hence be inert and also support a uniform flow of reactants to the seed. This demands efficient heat transport through the gas phase. Helium fulfils these requirements, as it is a noble gas with high thermal conductivity and low density. Growth rates of between 0.5 and 0.8 mm/h have been obtained at a sufficiently high precursor concentration, a reactor pressure of 200–800 mbar, and a seed temperature of 2000–2300 °C.

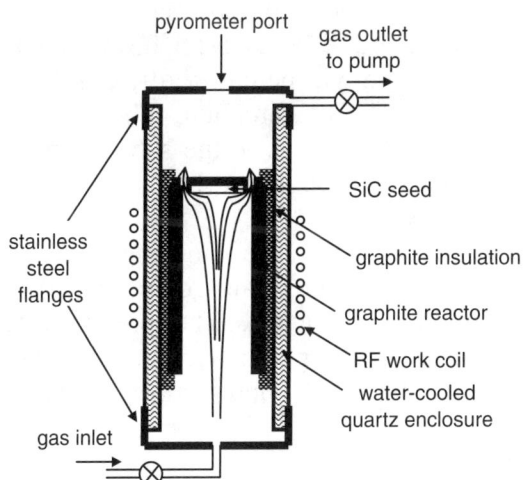

FIGURE 2.12   A schematic of a reactor for high-temperature chemical vapour deposition.

The standard cubic centimetre per minute (sccm) measures the volume flow of a gas at atmospheric pressure and a temperature of 0 °C.

At the high concentration of reactants used, with typical SiH_4 precursor flows in the range from 10 up to 100 standard cubic centimetres per minute (sccm), the C : Si ratio has to be chosen with care in order to avoid gas-phase nucleation, but still maintain a high growth rate. The C : Si ratio strongly depends on the reactor design, and values between 0.2 and 0.8 have been suggested. By keeping the carrier-gas flow rate relatively low, an efficient gas-phase cracking and limited cooling of the seed is obtained.

An obvious advantage with the HTCVD technique, is that most basic growth parameters, such as growth temperature, choice of carrier gas, supersaturation, precursor purity, C : Si ratio, and even the growth chemistry, can be chosen independently. This gives the possibility to adjust and accurately control the process within a wide parameter space, which points to the possibilities of making more thorough optimization than is possible in the other techniques. Some promising growth results have been shown, including micropipe densities of the order of below $100\,cm^{-2}$, and background doping levels below $10^{16}\,cm^{-3}$. The existing impurities are mainly Ti and B, most probably emanating from the graphite in the reactor enclosure [12].

2.2.2.7 Wafer fabrication from bulk crystals

The bulk crystals are preferably grown in cylindrical boules along the crystallographic c-axis. Before further wafer processing starts, the crystallographic orientations within the boule are accurately determined by X-ray diffraction, the $\langle 11\bar{2}0 \rangle$ and $\langle 1\bar{1}00 \rangle$ directions are marked, and the primary and secondary wafer flats are cut along those directions (FIGURE 2.13).

The boule is then sliced into substrates by using a diamond saw. If the substrates are to be used for further epitaxial growth of SiC, the 4H polytype will be cut 8° off, and the 6H polytype 3.5° off toward the $\langle 11\bar{2}0 \rangle$ direction. This will considerably reduce the formation of defects in the subsequently grown epitaxial layers. Substrates that should be used for other purposes are usually cut on-axis.

Due to the hardness of SiC, the polishing of the substrate surface is extremely demanding, and only recently have high-quality surfaces been produced. The preferred technique is a chemomechanical polishing, e.g. using a colloidal silica slurry at elevated temperature (55 °C) and at a pH > 10 [15].

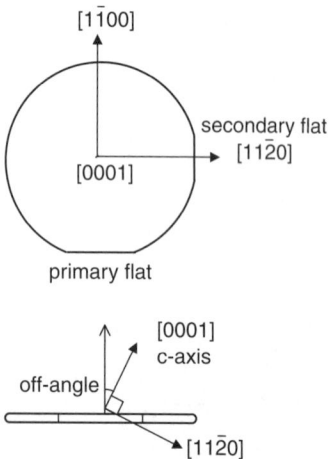

FIGURE 2.13 A substrate with indication of crystallographic orientations, main and secondary flats, and off-cut direction.

2.2.3 Applicability

Hardly any substrates used for semiconductor processing are currently produced by the early developed Acheson and Lely growth techniques. Instead, the Acheson technique is scaled up for production of large quantities of SiC for abrasive purposes, and material can, if the purity requirements are fulfilled, be used as a source powder for seeded sublimation growth. Until recently the Lely technique was regarded as being superior to the seeded sublimation growth, as the crystal platelets showed a lower micropipe density, even though the substrate size was limited to around one cm^2, but today substrates with equally large micropipe-free areas are produced by seeded sublimation.

The seeded sublimation growth technique is primarily suitable for the production of SiC bulk crystals and substrates of the 6H and 4H polytypes, currently up to a diameter of 10 cm, while larger diameters certainly will be reached in the near future. Some efforts have been made to also produce bulk crystals of the 3C polytype by this method, but no breakthrough has yet been reported. This might be due to the comparatively high temperatures needed for the sublimation, and that the 3C polytype is thermodynamically stable at lower temperatures. Growth at a lower seed temperature could hence be a solution.

The seeded sublimation growth technique could be modified by reducing the distance between the source and seed down to a few mm, and hence reducing the influence from the crucible walls and gaining better control of the growth parameters. This will reduce the defect formation, but also reduce the growth rate to below 100 μm/h [6] due to the reduced temperature difference between source and seed. This method, usually called close-space sublimation or sandwich sublimation, is suitable for fast growth of thick epitaxial layers (up to a few hundred micrometres) with a quality good enough for device applications. However, the growth rate is too low for bulk crystal growth.

Liquid phase epitaxy is still a rather immature technique, but has a large potential when the technological issues are solved. The typical growth rate reaches 100 μm/h, which is suitable for epitaxial growth of thick layers, but not enough for bulk crystals. The low defect generation probability and the possibility to close micropipes are probably the most important features of the LPE technology, and could be strong arguments to continue the development.

By using high-temperature chemical vapour deposition bulk crystals of several mm thickness have been grown. The crystal quality is comparable to the quality of crystals grown by seeded sublimation, but the growth rate is still below 1 mm/h. The advantage of using high-purity gas-phase precursors in a process, which is comparatively easy to control, facilitates fabrication of material with extremely low background levels of impurities.

This makes HTCVD suitable for growth of bulk crystals of the 6H and 4H polytypes for highly specialised applications where the substrate purity is of great importance. HTCVD is also well suited for thick, low-doped epitaxial layers on conventional substrates.

2.2.4   Comparison of techniques

Some important aspects to evaluate the growth techniques for SiC bulk crystal growth are given in TABLE 2.2. The ranking is made with + and − signs, showing the advantages and disadvantages with the different techniques. The ranking is based on published results, rather than on estimations of future possibilities. The LPE and HTCVD techniques are under rapid development, and it is still difficult to predict their final potential.

The growth rate is a key parameter for the time efficiency of the process, and should be at least 1 mm/h in order to be suitable for bulk crystal fabrication, and ideally at least 10 mm/h. (The typical Si bulk growth rate using the Czochralski technique is above 100 mm/h.) Usually there is a trade-off between the growth rate and crystal quality, as sacrificing the quality can increase the growth rate in most cases. While maintaining a crystal quality suitable for device substrates, growth rates of 1 mm/h or above have been reported for seeded sublimation growth. This is

TABLE 2.2   Comparison of seeded sublimation, LPE and HTCVD growth of bulk SiC crystals.

Technique Parameter	Seeded sublimation	LPE	HTCVD
Growth rate	+	−	−
Boule diameter	+ +	+	+
Boule length	+ +	−	+
Defect control	+	+ + +	+
Impurities	+	+	+ + +
Stability/control	+	+	+ +
Equipment cost	+	− −	−
Process cost	+ +	+ +	−
Maturity	+ + +	− −	−
Potential	+ +	+	+ +

acceptable, but there is still room for improvement. Both LPE and HTCVD show insufficient growth rates of below 1 mm/h at a reasonable crystal quality.

The length and diameter of the boule also determine the process suitability for bulk crystal production. The boule diameter gives the largest substrate diameter, and the boule length determines the number of wafers grown in one run of the reactor. Both parameters are mainly related to the maturity of the technique, and to the effort spent on up-scaling the processes, which makes the seeded sublimation the most successful technique today in both these respects. The boule diameter is mainly limited by the radial uniformity of temperature and source supply in the process chamber. In practice, all the techniques should have the potential to reach a diameter of 10 cm, and beyond. The importance of the reactions at the crucible walls in the HTCVD process might be limiting for scaling it to even larger diameters. The length of the boule is determined by the process stability over time. In the seeded sublimation and LPE the change in stoichiometry of the source due to depletion of reactants will limit the boule length, while no such limitations are expected in the HTCVD.

The control of defects and incorporation of impurities is crucial for the use of the material for further device processing. The defect generation is due to the process stability and can be a trade-off with the growth rate. It has proven difficult to reduce the inclusions of foreign polytypes and the micropipe concentration in both seeded sublimation and HTCVD growth, while the LPE growth has been shown to be much more forgiving. This strongly favours LPE in this respect. The impurity incorporation is mainly determined by the quality of the source material and the purity of the material used for the heated parts of the crucible and reactor enclosure. The possibility to use high-purity source gases and high-purity SiC-coated graphite parts in the reactor for HTCVD gives it an advantage in high-purity growth. Both in seeded sublimation and in LPE the crucible will evaporate during the process and impurities will be released. It is also difficult to find solid SiC sources of the same high purity as semiconductor grade gases.

The total substrate production cost can be divided into equipment cost and process cost, including precursors and service. The equipment cost is mainly determined by the complexity of the system, and this is of course related to the process performance, which improves with advanced process control, choice of materials in critical reactor parts and general robustness. Given a similar standard of the equipment, the seeded sublimation would probably have the simplest solution, while HTCVD will need a high-performance gas-delivery system, including safety precautions for handling pyrophoric and toxic gases at high temperatures, and the LPE process probably needs an even more complex reactor, handling a combination of high pressure and extremely high temperatures. The LPE furnace might hence be the most difficult process equipment to scale up to larger diameters and boule lengths. The process cost is mainly determined by the precursor cost, as a similar amount of maintenance for the different systems can be assumed. Hence, the gases used in the HTCVD are more expensive than the solid source material for sublimation and LPE, even though the cost to a large extent is determined by the required material purity.

Among the techniques to be considered for bulk growth of SiC, seeded sublimation is by far the most mature, and also the one used for production of substrates. Hence, most process development is made in this technology and the process results are continuously improved. Only a few groups work on LPE and HTCVD, and the development of these processes to a level comparable to the seeded sublimation is far ahead.

Nevertheless, the potential of the HTCVD and LPE, given an appropriate development effort, is great. The potential of the HTCVD is mainly related to the availability of high-purity source and reactor materials, and the independent selection of process parameters, which facilitate an extensive optimization. The LPE process still needs much development, and the major driving force will be improved defect control.

2.2.5   Overcoming problems

2.2.5.1   *Seeded sublimation growth*

Despite a major development effort on seeded sublimation growth, process limitations still remain. These are related to the impurity and defect concentration of the crystals, the non-uniformity within the crystal boule (both axially and radially) and the stress within the produced substrates (increasing with the boule diameter).

Improved process control is the single most important step to overcome these problems. Hence, the temperature distribution should be radially homogeneous, and the axial gradient well controlled in the crucible. This will reduce the thermal stress in the crystal, which causes cracks, and also provide better control of the vapour-phase composition and nucleation on the growing surface, which reduces the probability of inclusions of graphite or Si droplets. Also, a stable pressure and ambient control is of utmost importance, as particles from the source can be transported by convection to the growing surface under non-optimum conditions. Particle generation is also less severe from a solid polycrystalline source than from a powder source. The impurity incorporation can be controlled by an appropriate choice of crucible material, and in addition to the conventional graphite and SiC-coated graphite parts, a tantalum crucible has been suggested. By using a Ta crucible, excess carbon and impurities released from the crucible can be eliminated, and the graphitization of the source can be significantly suppressed. Ta getters C by forming TaC, which reduces the formation of particles and graphite inclusions in the grown material [6].

The level of impurities obtained in seeded sublimation growth, can be determined by the purity of the ambient gas, of the source and of the crucible material. The purity of the ambient gas and the source are related to the purification of raw materials used for fabrication, and is improved to a very high degree. Rather, the limiting function is the impurities in the crucible material, both elemental impurities, such as Al, B and Ti, and gas impurities, such as nitrogen, oxygen and humidity. The gases can to some extent be released by a high-temperature de-gassing step, while the atomic impurities are extremely difficult to eliminate. A SiC coating of the crucible or a crucible made entirely of polycrystalline SiC is an attractive solution, with the drawback that the crucible material will evaporate and be consumed during the sublimation process. Ta crucibles can be available in high-purity material, with the drawback that Ta concentrations of up to $10^{16} \, \mathrm{cm}^{-3}$ can be obtained in the bulk SiC [6].

Polytype control is an important issue in bulk growth, related to the small difference in formation energies between the polytypes in SiC, and the fact that the (0001) lattice planes, on which most growth is performed, do not contain the polytype information. The face of the seed and the growth conditions are usually more important parameters to control the polytype of the grown crystal, than the polytype of the seed itself. A way to force the grown crystal to maintain the polytype of the seed is to reveal the full stacking sequence by using a slightly off-cut seed (3–10°) where a cross section of the lattice appears on the surface. Growth on the (1$\bar{1}$00) or (11$\bar{2}$0) lattice planes has also proven successful, as the polytype of the boule perfectly matches the seed (FIGURE 2.14). This process also has the advantage that micropipes do not propagate perpendicular to the c-axis, and hence the crystals are free from micropipes [13]. The main drawback is that the preferred (0001) oriented substrates must be cut perpendicular to the boule, with a loss of material.

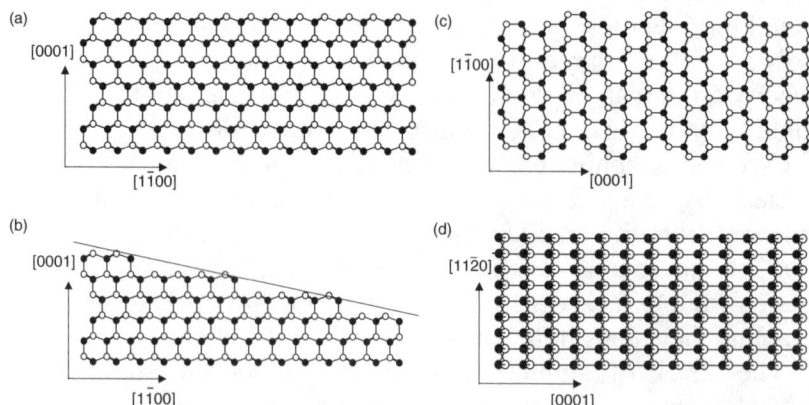

FIGURE 2.14   Projections of the 4H SiC lattice: (a) on-axis (0001) plane; (b) off-axis (0001) plane; (c) on-axis (1$\bar{1}$00) plane; (d) on-axis (11$\bar{2}$0) plane.

2.2.5.2   *Liquid phase epitaxy*

Equipment development for LPE is a major problem, and could delay the realisation of a commercial process. The high reactivity of the Si melt puts high demands on the crucible material, and until now the obvious choice has been graphite, which is dissolved in the melt and participates in the growth process. However, the impurities in the graphite are dissolved as well, and are incorporated into the grown crystal. As molten Si also attacks refractive metals, no stable material for the crucible has yet been reported. Graphite, or polycrystalline SiC of an extremely high purity, would be possible to use, but the available grades are not yet sufficient. In addition, the combination of high temperature, high pressure, substrate rotation and pulling translation in an environment that has to be extremely well controlled, is an engineering challenge. An alternative, for small substrates, is to use a crucible-free technique by applying electromagnetic levitation, thus keeping the melt and seed away from the reactor walls.

A serious problem in LPE growth is the possible incorporation of particles, solvent droplets or gas bubbles into the growing crystal. These defects emanate from a roughness of the growing surface, forming a hollow structure, which is subsequently overgrown. This can be avoided by maintaining an atomically flat interface between solid and liquid during growth. One requirement is an accurate control of the supersaturation, source flow and temperature at the growing crystal, which can be achieved by substrate rotation, and an advanced system for temperature regulation. The possibility to adjust the composition of the melt during growth would also be beneficial.

In general, the success of LPE growth of SiC strictly relies on a massive development effort, comparable to that made in seeded sublimation growth.

2.2.5.3 *High-temperature chemical vapour deposition*

The simultaneous development of process and equipment as one structure is fundamental for the success of the HTCVD technology. At the process temperatures, up to 2300°C, the heated graphite reactor parts are easily decomposed, especially if hydrogen is present in the ambient. Hence, H_2 must be excluded from the carrier gas, and only a noble gas can be used. In addition, the carrier gas must have a low density and a high heat conductivity, in order to efficiently transfer heat to the precursors in the gas phase of the cracking zone and support the formation of reactive Si-C radicals in the gas phase. This favours He as the carrier gas, but excludes Ar. However, as long as the precursors contain hydrogen, the graphite reactor walls will deliver carbon to the growth. Also, Si can attack the walls, forming SiC_2 and Si_2C [14]. The use of coated graphite (e.g. by SiC) in the heated parts of the reactor could be a solution to these problems. Hence, the temperature adjustment must be made to lower temperatures to prolong the lifetime of the coating.

Si droplets or SiH_2-based polymers are known to nucleate from SiH_4 in the gas phase, forming a 'cloud' at the reactor inlet. If the SiH_4 flow is low and the temperature is high in the zone, this cloud will evaporate while passing the cracking zone. In the presence of hydrocarbons more stable clusters, containing Si-C bonds, will form and the gas phase could be depleted of reactants. By maintaining a high enough temperature in the cracking zone gas phase, the clusters will evaporate and participate in the nucleation process at the seed. If, in addition, the seed temperature is low enough to keep a supersaturation at the growing surface, a stable growth regime is found, where a high growth rate can be obtained. If the temperature profile is uneven, with cold fingers reaching the seed, Si droplets or Si-C clusters may nucleate on the surface, which will result in crystal defects and micropipes [14].

The gas-phase nucleation of Si-C clusters increases with the increase of the C:Si ratio in the gas phase, which might result in a lowering of the growth rate. This can be solved by a high temperature, but also by reducing the reactor pressure, which lowers the probability of forming clusters. Pressures between 200 and 800 mbar are hence optimal.

Polytype control is obtained by the choice of seed crystal, as in seeded sublimation. Hence, the 6H polytype can be grown on both the (0001) Si and (000$\overline{1}$) C faces of 6H seeds, while 4H is preferably grown on the (0001) C face of 4H seeds. As growth on off-oriented seeds will provide a cut through the stacking sequence of the lattice, polytype control will be easier and island nucleation avoided. 4H–6H polytype transformations can occur if the temperature suddenly drops, or at low gas phase C:Si ratios [14].

2.3 EPITAXIAL GROWTH

2.3.1 Introduction

Thin film growth by vapour phase epitaxy (VPE) has matured during the last decade, and today commercial reactors for SiC growth are available from at least three different suppliers. The largest can handle up to seven 2-inch wafers. The reactor designs are based on concepts developed for growth of III–V compound semiconductors, but adapted for temperatures up to 1600 °C, rather than 700 °C used for the III–V compounds. The reactors support substrate rotation for uniformity of doping and thickness, and advanced gas-handling systems for abrupt doping interfaces and accurate process control.

The basic principle for VPE is to feed precursor gases diluted in a carrier gas into a reaction chamber, where growth takes place on a heated seed crystal. The main process parameters can be independently controlled over a wide range and optimized to improve the crystal quality and to build the intended device structure.

Two major improvements of the epitaxial process have been important to enable growth of device structures, where doping control and low defect concentration are fundamental. The polytype control and defect reduction rely on step-flow epitaxy, where growth is performed on an off-cut substrate, where the lattice structure is revealed, and inclusion of foreign polytypes is reduced. In site-competition epitaxy the fact that dopants are incorporated on either Si or C sites in the lattice has been used, and by tuning the C:Si ratio in the gas phase, the doping incorporation can be controlled.

2.3.2 Tutorial

2.3.2.1 Reactor design

Reactors for growth of thin epitaxial layers for device applications should be designed with the demands on layer uniformity and process control in mind. The growth rate is limited by the diffusion of reactants to the growing surface. In a simple growth model a thin boundary layer above the substrate is depleted of reactants, and the gas phase above this layer maintains a constant precursor concentration. The thickness of the depleted layer increases along the susceptor in the gas flow direction. At a higher carrier gas flow rate, the reactants will be preserved for deposition downstream of the susceptor, which will improve the uniformity at the cost of lower precursor utilisation. The reactor pressure will not influence the gas-phase depletion, but a low pressure, together with a high carrier-gas flow velocity and a low reactor-cell height, is important to reduce the buoyancy-driven convection. This is mainly due to temperature differences in the gas phase, i.e. differences in gas density. Convection can reduce the layer uniformity and contribute to dopant carry-over from one grown layer to the next, so-called memory effects.

A consideration especially important in the design of reactors for epitaxial growth of SiC is the choice of material in the heated parts and thermal insulation of the reactor. Growth is usually performed at temperatures between 1450 and 1650 °C, which only a few materials can sustain. In this temperature range also radiation dominates over convection as the major mechanism for temperature loss. This makes it important to choose the correct materials, and to reduce the radiation by efficient shielding.

Small reactors, handling substrates up to a few cm^2, can easily be designed with a water-cooled quartz enclosure around a graphite susceptor heated by radio frequency (RF) induction (FIGURE 2.15) [16]. The thermal insulation from the quartz can possibly be made by placing porous graphite felt below the susceptor, and the radiation from the small top surface can be tolerated. Reactors for larger substrates require a more sophisticated solution, with a heat shield surrounding the whole

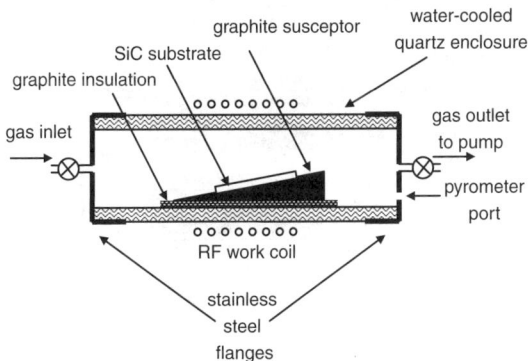

FIGURE 2.15   A schematic of a small horizontal reactor.

susceptor, e.g. by graphite felt or a quartz enclosure coated with a thin, highly reflecting film of gold or graphite. It is also important to add an insert, which guides the gas flow over the substrate in order to improve layer uniformity (FIGURE 2.16) [17].

A related solution is the hot-wall reactor, produced by Epigress [18]. This reactor features a graphite susceptor tube, where the ceiling, the walls and the bottom plate with the substrate are all heated by RF induction. The susceptor tube, with surrounding insulation of graphite felt or porous graphite, is fixed as a plug in the quartz enclosure and designed for optimum uniformity (FIGURE 2.17) [19]. In order to further improve the thickness uniformity, the reactor can be supplied with a rotating substrate holder.

The main drawback with the horizontal reactor is the limited possibility to scale it up for simultaneous growth on multiple substrates, as efficient scaling up needs a design based on rotational symmetry. Two different solutions have been presented, and both are commercially available.

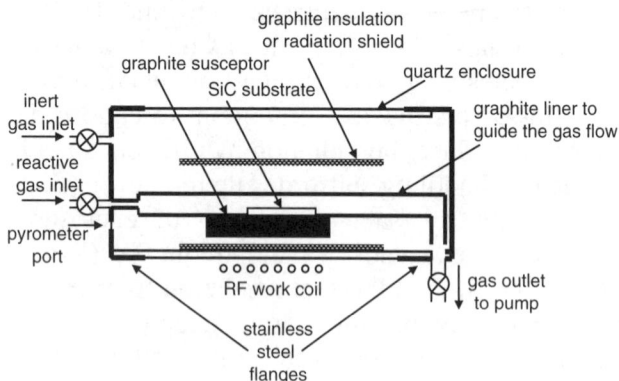

FIGURE 2.16   A schematic of a horizontal reactor with gas liner.

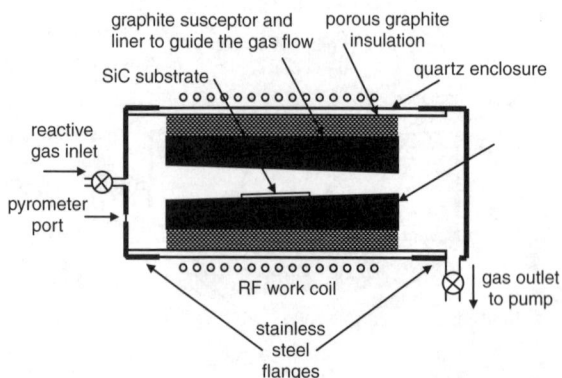

FIGURE 2.17   A schematic of a horizontal hot-wall reactor.

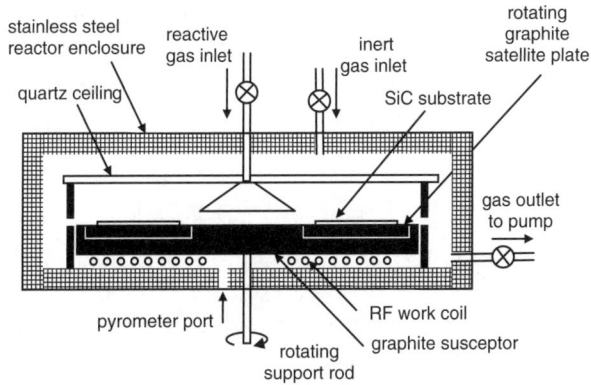

FIGURE 2.18   A schematic of a planetary multi-wafer reactor.

The solution most similar to the horizontal reactor is the planetary reactor developed by Aixtron, and produced by Epigress [18]. This reactor has an inlet for the active gases at the centre of a rotating graphite plate susceptor, and the gas flows parallel to the plate. The substrates are placed on individually rotating graphite satellites, inserted into the large plate, and a pancake RF coil heats the susceptor. The reactor ceiling is heated by radiation from the susceptor and adjusting the forced cooling of the ceiling can control the temperature (FIGURE 2.18) [20]. The satellite is rotated by a gas-foil flow introduced below the satellite plate to a speed of about 50 rpm and the large plate is mechanically rotated at about 10–20 rpm. The total diameter of the susceptor is 300 mm and it can simultaneously handle seven 50-mm diameter substrates.

A different solution is the vertical reactor, produced by Emcore [21]. Here, the gas is supplied perpendicular to the susceptor, through an inlet flange, which allows accurate control of the radial gas distribution. The inlet and the reactor walls are water-cooled stainless steel. The distance between the gas inlet and the susceptor is about 150 mm and in order to avoid convection in the comparatively large enclosure with large temperature gradients, the carrier gas flow has to be high, and the rotation speed of the susceptor between 500 and 1000 rpm. This reactor concept has been used both for a 75-mm diameter graphite susceptor for one single 50-mm substrate, [22], and for a 180-mm diameter susceptor for six 50-mm substrates. In the multi-wafer design the susceptor consists of a molybdenum platen, on which a graphite plate with wafer pockets is placed (FIGURE 2.19) [23].

The gas-delivery system for the reactors is computer controlled. The gas flow is directed by pneumatic valves and regulated by mass flow controllers. Dopant delivery is made through a dilution network, which allows control over 3–4 orders of magnitude.

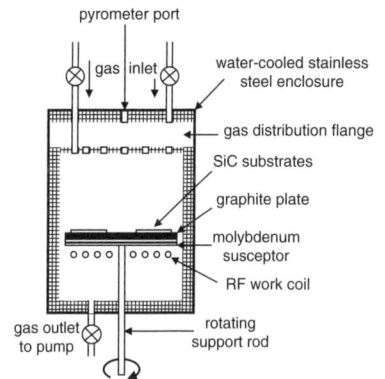

FIGURE 2.19   Schematic of a vertical multi-wafer reactor.

A throttle valve placed in the pump line regulates the reactor pressure. A burner or wet scrubber placed after the pump can handle the gas waste of non-reacted precursors and polymers and other by-products produced in the process.

2.3.2.2   The epitaxial growth process

SiC is most commonly grown using silane, SiH_4, and propane, C_3H_8, as the main growth precursors for Si and C, respectively. Alternatives to silane are, for example, disilane and chlorosilanes, and to propane other hydrocarbons, for example, ethylene and methane. For special demands, such as low-temperature processes, precursors combining both carbon and silicon in one molecule have been used.

The growth temperature should be chosen high enough to decompose the precursors, which in hydrogen requires about 1200 °C for silane and 800 °C for propane. The temperature should also be high enough to support the necessary chemical surface reactions and to allow the surface mobility of the growth species, and ultimately to provide the formation energy for the desired SiC polytype to grow [24]. This usually means a temperature between 1450 and 1650 °C, where 4H is grown at the higher, and 6H at the lower temperatures in the range.

Hydrogen is the preferred carrier gas. It is comparatively inexpensive and can be obtained at an extremely high purity by using a heated palladium membrane purifier at the point of use. Hydrogen also has a high thermal conductivity, low viscosity, and low density, which will provide a stable laminar flow under most conditions. Hydrogen will partly decompose at the growth temperature, and the atomic hydrogen will participate in the growth process as a light etchant on the growing surface, and hence increase the surface mobility of reactants, thus reducing the formation of inclusions and defects. A drawback is that atomic hydrogen will attack the reactor surfaces, especially graphite (by forming methane), but also any SiC coating. In horizontal reactors, and at low growth temperatures, a mixture of hydrogen with 10–50% argon has been shown to improve the crystal quality and the growth uniformity, while the growth rate decreased. This is most probably due to a reduced diffusivity of reactants through the carrier gas, combined with a reduced cooling of the growing surface [25].

The flow rates depend on the reactor geometry. The carrier-gas flow rate typically has to increase with the reactor cross section, and flow rates between 3 and 80 standard litres per minute are used. The lower flows are used in small horizontal reactors and the highest in the vertical multi-wafer reactor. In the preferred growth regime, at typical growth rates of 2–8 μm/h, the growth rate is proportional to the silane flow. The molar fraction of silane in the carrier gas can hence be around 1×10^{-3}.

The reactor pressure is usually a non-critical growth parameter, and pressures from atmospheric down to 50 mbar are used. The main advantage with reduced pressure is an improved laminar gas-flow profile, which provides stable and uniform growth.

The C : Si ratio is the ratio between C and Si atoms in the gas phase. This ratio is dependent on the growth parameters, and it can be reduced to the limit where Si droplets appear on the surface. This limit occurs at a low C : Si ratio of below 1.0, down to 0.5 at low reactor pressures and temperatures, while at atmospheric pressure and high temperatures the C : Si ratio must be increased to 1.5 or 2.0. This can partly be explained by the difference in cracking temperature for silane and propane, which gives a higher efficient C : Si ratio on the growing surface than in the gas phase, especially at lower temperatures. Too high a C : Si ratio gives a high supersaturation, which limits the surface mobility of growth species and supports island nucleation. This will appear as a rough surface and an increased number of defects. The C : Si ratio only has a limited effect on the growth rate, but will strongly influence the doping incorporation.

A typical growth sequence starts with etching the substrate for a few minutes in order to remove both the native oxide and remaining polishing damage as well as to provide a smooth surface for subsequent growth. The most aggressive etchant in use is hydrochloric acid, HCl. Diluted in the hydrogen carrier gas, at a temperature of between 1300 and 1500 °C a few minutes of etching is sufficient to remove the damage. Also, pure hydrogen can be used. This etchant needs higher temperatures, up to the growth temperature, and a somewhat longer etching time. A way to further reduce the etch rate is to mix propane with hydrogen. This mixture is less aggressive to the reactor enclosure, and is less likely to produce silicon droplets at the substrate surface, although 15–30 min of etching time at the growth temperature can be needed to obtain a smooth surface. After the etching, the growth temperature is adjusted and growth starts, as the precursors are mixed into the carrier gas.

Overall reaction for silane and propane forming SiC at growth [24]:

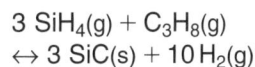

$$3\,SiH_4(g) + C_3H_8(g) \leftrightarrow 3\,SiC(s) + 10\,H_2(g)$$

Some of the gas phase reactions:

$$SiH_4(g) \leftrightarrow SiH_2(g) + H_2(g)$$
$$Si(g) + H_2(g) \leftrightarrow SiH_2(g)$$
$$SiH_2(g) + Si(g,l) \leftrightarrow Si_2H_2(g)$$
$$Si_2(g) + H_2(g) \leftrightarrow Si_2H_2(g)$$
$$C_3H_8(g) \leftrightarrow CH_3(g) + C_2H_5(g)$$
$$CH_3(g) + H_2(g) \leftrightarrow CH_4(g) + H\,(g)$$
$$2\,C_2H_5(g) \leftrightarrow 2\,C_2H_4(g) + H_2(g)$$
$$C_2H_4(g) \leftrightarrow C_2H_2(g) + H_2(g)$$

Some of the surface reactions:

$$Si(g) \leftrightarrow Si(s)$$
$$SiH_2(g) \leftrightarrow Si(s) + H_2(g)$$
$$Si_2H_2(g) \leftrightarrow 2\,Si(s) + H_2(g)$$
$$C_2H_4(g) \leftrightarrow 2\,C(s) + 2\,H_2(g)$$
$$C_2H_2(g) \leftrightarrow 2\,C(s) + H_2(g)$$
$$CH_4(g) \leftrightarrow C(s) + 2\,H_2(g)$$

2.3.2.3 Dopant incorporation

The dopants used in SiC VPE are aluminium and boron for p-type and nitrogen and phosphorus for n-type, with trimethylaluminium, $(CH_3)_3Al$, diborane, B_2H_6, nitrogen gas, N_2, and phosphine, PH_3, as the most common dopant precursors. The precursors are all gases, except trimethylaluminium, which is a liquid, and delivered into the reactor from a bubbler by using, for example, hydrogen as a carrier gas.

As p-type dopant, aluminium is the most frequently used. It has a low diffusivity in SiC over the temperature range of interest, high incorporation efficiency and, compared to boron, low acceptor ionisation energy. The incorporation is almost linear to the flow of trimethylaluminium and a maximum doping level of about $2 \times 10^{20}\,\mathrm{cm}^{-3}$ can be reached. Boron is less used, partly due to its higher acceptor ionisation energy and lower solubility in SiC. Boron also has a higher diffusivity than aluminium, and a tendency to stick to the reactor walls, and subsequently evaporate. Both these effects lead to non-abrupt doping profiles. The boron doping incorporation saturates around $1 \times 10^{20}\,\mathrm{cm}^{-3}$ [26].

As n-type dopant, nitrogen is preferred to phosphorus, even though the acceptor ionisation energy is practically the same for both. The incorporation of nitrogen from N_2 is a thermally limited process, even at the growth temperatures, due to the high dissociation energy of the gas. The dopant incorporation is proportional to the N_2 flow, and saturates at around $1 \times 10^{20}\,\mathrm{cm}^{-3}$. The vapour pressure of phosphorus is extremely high at the growth temperature, which reduces the incorporation probability two orders of magnitude below that of nitrogen. In addition, phosphine is a highly toxic precursor, and should thus be avoided, if possible [27].

The dopants are incorporated into a Si or a C lattice site. A large dopant atom will preferably replace the larger Si atom in order not to distort the lattice. Aluminium and phosphorus, both with an equivalent covalent radius comparable to that of silicon (TABLE 2.3) are hence regularly incorporated into Si lattice sites. Dopants with smaller diameters can be incorporated into either lattice site, without disturbing the lattice, which is the case for nitrogen. On the other hand, boron seems to be incorporated into the Si site only. This can be understood from the fact that hydrogen is incorporated together with boron, forming a B-H complex, too large to fit into a C site [27].

By adjusting the C : Si ratio in the gas phase, and hence on the growing surface, the incorporation of the dopant atoms can be controlled. At a high C : Si ratio the C lattice sites are readily occupied by carbon. Incorporation of dopants, which preferably occupy the C site, will be hindered, while incorporation of dopants into the Si sites will be facilitated due to the shortage of silicon. Hence, incorporation of aluminium, phosphorus and boron (as the B-H complex) will be enhanced at a high C : Si ratio, while the nitrogen incorporation will be limited. On the other hand, at a low C : Si ratio, the nitrogen incorporation will be enhanced, and incorporation of the other dopants is limited. This

TABLE 2.3 Non-polar covalent radii for some elements and the equivalent size of the B-H compound [27].

Species	Radius (Å)
Si	1.17
C	0.77
P	1.10
N	0.74
Al	1.26
B	0.82
B-H	1.10

explanation holds for growth on the (0001) Si face, and for doping with boron, aluminium and phosphorus for growth on the (000$\bar{1}$) C face. Nitrogen incorporation for growth on the (000$\bar{1}$) C face reaches a maximum for medium C : Si ratios, but decreases for both low and high C : Si ratios (TABLE 2.4). This indicates that nitrogen can be incorporated into both C and Si lattice sites, and that Si lattice sites are preferred at a high, and C lattice sites at low, C : Si ratio, respectively [27]. In general, the influence on the dopant incorporation is most dramatic at the lower C : Si ratios, close to Si droplet formation at the growing surface, and saturates at higher ratios (FIGURE 2.20).

The hydrogen incorporated in the lattice together with boron has been found to passivate the boron acceptor. Probably this is achieved by hindering the transfer of an electron from a carbon, to the boron atom. A similar, but weaker, interaction has been observed between aluminium and hydrogen. By performing an anneal of the aluminium- or boron-doped sample in an inert atmosphere, e.g. argon at a temperature above 1000°C for 30 min, the hydrogen is released, and the net carrier concentration accordingly increases [28].

The lowest background doping levels obtained are in the low 10^{14} cm^{-3} range. n-Type doping (nitrogen from the gases) dominates in the hot-wall reactor [19] and p-type (aluminium and boron from uncoated graphite parts) in the vertical cold-wall reactor [22]. The difference in surface kinetics and precursor cracking could also influence the dopant incorporation.

2.3.2.4 Polytype control and defect reduction

Vapour phase epitaxy is performed at a relatively low temperature, where 3C is the thermodynamically stable polytype. This means that island nucleation of 3C SiC will occur on on-axis substrates, which provides no information about the lattice stacking sequence. Due to the two possible orientations of 3C crystals on a hexagonal substrate, double-positioning twinning occurs. On the Si face, double-positioning boundaries could be clearly seen between the domains of opposite twinning, while on the C face the surface becomes rough due to twinned micro-crystals.

In order to control the polytype, growth could be performed on off-axis substrates, revealing the full stacking sequence of the lattice, and providing a high density of step sites for nucleation. The terraces between the steps should be narrow as compared to the surface diffusion length of the growth species so that nucleation occurs at the steps, where the growth could follow the stacking sequence and hence the polytype of the substrate (FIGURE 2.21). This growth mode is called step flow growth [16].

TABLE 2.4 Summary of site-competition behaviour.

Parameter	C:Si	P	N	Al	B-H
Si face	⇑	⇑	⇓	⇑	⇑
	⇓	⇓	⇑	⇓	⇓
C face	⇑	⇑	⇓	⇑	⇑
	⇓	⇓	⇓	⇓	⇓

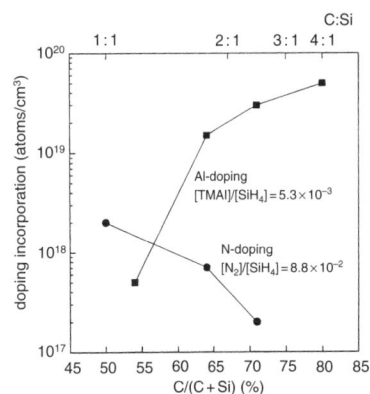

FIGURE 2.20 Nitrogen and aluminium incorporation from N$_2$ and TMAl in a horizontal reactor. In order to avoid infinite scales, the dopant incorporation is given as a function of the normalised C/(C + Si) ratio. From [17].

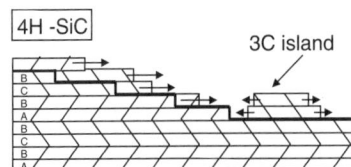

FIGURE 2.21 An illustration of the principle for step flow growth, and the formation of 3C inclusions.

TABLE 2.5 Step height on off-cut substrates, in number of Si-C bilayers at highest and next highest probability [16].

Polytype/ face	Step height ≥70% of the bilayers	Step height ≈20% of the bilayers
6H/Si	3	–
6H/C	1	3
4H/Si	4	2
4H/C	1	2

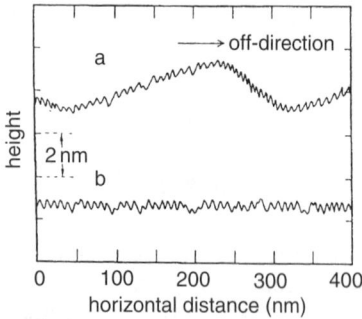

FIGURE 2.22 Step bunching on 6H 5° off-axis substrates. From [16]. (a) (0001) Si face; (b) (0001̄) C face.

Off-orientation towards the <112̄0> direction is preferred. An angle of 3.5° is employed for 6H, while an 8° off-angle is needed for 4H, which has a larger terrace width for the same off-angle and smaller stacking fault energy than 6H. A smaller off-angle can be compensated for by an increased surface mobility, either by raising the growth temperature or by reducing the supersaturation at the growing surface (i.e. lower the growth rate or the C : Si ratio).

The typical step height on an off-cut substrate after growth depends on the polytype and the face. For example, more than 90% of the steps on a 6H (0001)Si face substrate consist of three Si-C bilayers (TABLE 2.5), where each bilayer is 2.5 Å high. The number of bilayers in each step reflects the lowest surface energy for the particular surface. Under epitaxial growth, the steps on the (0001) Si face have a tendency to bunch and form macrosteps, with a height of a few nanometres. On 6H, the 3–6 nm high macrosteps are composed of several microsteps with different step height (FIGURE 2.22), while 4H produces genuine macrosteps of 10–15 nm height. On the C face for both polytypes, no macrosteps are observed [16]. Under non-optimum growth parameters, i.e. at high supersaturation and low temperatures, the step bunching becomes more enhanced, and step heights of 30–80 nm can occur.

A stable step flow is crucial for defect-free growth. A point defect on the surface, e.g. a micropipe, a silicon droplet, a scratch or a particle, can easily hinder the step flow, and a basal plane terrace will grow behind the obstacle (FIGURE 2.23) [20]. The

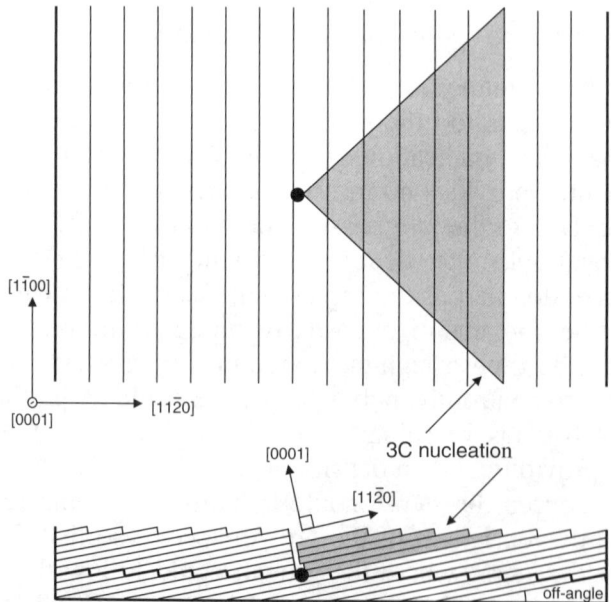

FIGURE 2.23 Mechanism of defect formation on an off-axis substrate.

terrace will form an inclined bottom of a pit in the surface. If the pit is shallow, it will be filled and form an amphitheatre-shaped defect. If the terrace grows large as compared to the surface diffusion length of the growth species, it can be the site for 3C nucleation, which appears as an equilateral triangular defect, oriented with the apex in the original obstacle and the base in the step-flow direction on the surface.

Polishing scratches can easily be decorated by tetrahedral pits or other point defects. The scratches can also be seen as line defects on the surface after growth. As polishing damage can be hidden, and is not always seen in an interference contrast microscope, a thorough preparation of all wafers before growth, including in situ etching, is crucial to avoid these defects [20].

2.3.2.5   Growth of 3C SiC on Si substrates

As an alternative to the homoepitaxial growth of the hexagonal 4H and 6H polytypes on SiC substrates, the cubic 3C polytype can be heteroepitaxially grown on Si substrates. The main problem in this process is the large lattice mismatch of 20% between Si and 3C SiC, which can be overcome by a carbonization process, which forms a SiC buffer layer before the growth starts.

In a conventional reactor for SiC growth, the Si substrate is heated rapidly (up to 10 K/s) to the growth temperature or even higher, but below the Si melting point of 1410 °C, under a flow of propane at a molar fraction of about 1×10^{-3} in the hydrogen carrier gas. The temperature is kept for a few minutes, while the substrate surface is carbonized, i.e., silicon reacts with carbon, forming a few nanometres thick SiC film. To start the growth, the susceptor temperature is stabilized at around 1250–1350 °C, the propane flow rate is lowered about an order of magnitude to the rate used for growth, and the silane flow is added (FIGURE 2.24). Growth rates of a few micrometres per hour can readily be obtained [29].

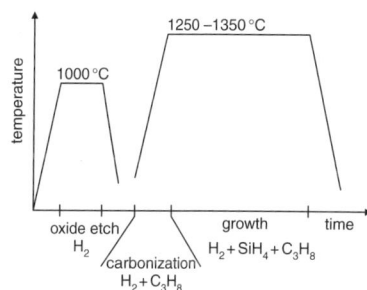

FIGURE 2.24   Growth sequence for 3C SiC growth on Si.

In an alternative process the growth is performed in a horizontal hot-wall low-pressure furnace, with a batch of silicon substrates placed vertically (FIGURE 2.25). The carbonization takes place at a pressure of below 0.1 mbar, under a partial pressure of 10% of acetylene, C_2H_2. The temperature ramping speed is about 10 K/min up to 1020 °C, where carbonization proceeds for one hour. This will form a SiC layer, a few nanometres thick. The growth is then performed at 750–1000 °C, under an alternating supply of acetylene (for 10 s) and dichlorosilane, SiH_2Cl_2, (for 20 s) at a growth rate of about one nanometre per cycle. The efficiency of this process is given by the number of

39

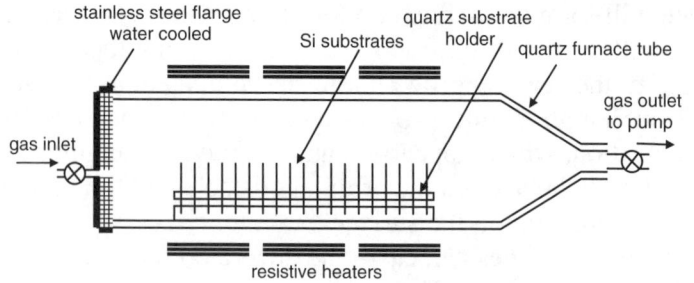

FIGURE 2.25   A schematic of a horizontal hot-wall reactor for growth of 3C SiC on Si substrates.

simultaneously processed wafers, and the excellent thickness uniformity of ±0.5% over a 15-cm diameter substrate [30].

The parameters to be optimized for formation of a high-quality buffer layer are the ramping speed, the carbon precursor flow rate, the carbonization temperature and the time. The main problem to be avoided is the formation of voids in the substrate, which occur when the supply of carbon at the surface is insufficient to form a protecting SiC layer during the temperature ramp. Increasing the final temperature or carbonization time will increase the thickness of the buffer layer.

As the SiC buffer layer nucleates in islands, and the monoatomic substrate does not contain any information about the preferred stacking directions of SiC, the islands will occur in twin relations to each other. Anti-phase boundaries will be present between regions separated by an odd-atom step in the substrate (FIGURE 2.26), but they usually annihilate as growth proceeds. The occurrence of both the twinning and anti-phase regions can be reduced by growing on off-cut silicon substrates. The crystallinity of the 3C layer can also be improved by optimizing the C : Si ratio in the gas phase, and keeping in mind that a high growth rate usually needs a high growth temperature.

2.3.3   Applicability

The epitaxial techniques are ideal for the growth of thin layers of high quality for device applications. With typical growth rates of 2–8 μm/h, layers up to a thickness of 50 μm are within reach. Thicker layers should preferably be grown by a modified bulk growth technique, e.g. close-space sublimation or high-temperature CVD, which on the other hand are less suited for growth of thin layers with controlled and variable doping. A chimney reactor operating at the intermediate temperature of 1850 °C, and reaching a growth rate of above 20 μm/h has also been suggested, and could be suitable for growth of thick layers with some control [12].

(a)

(b)

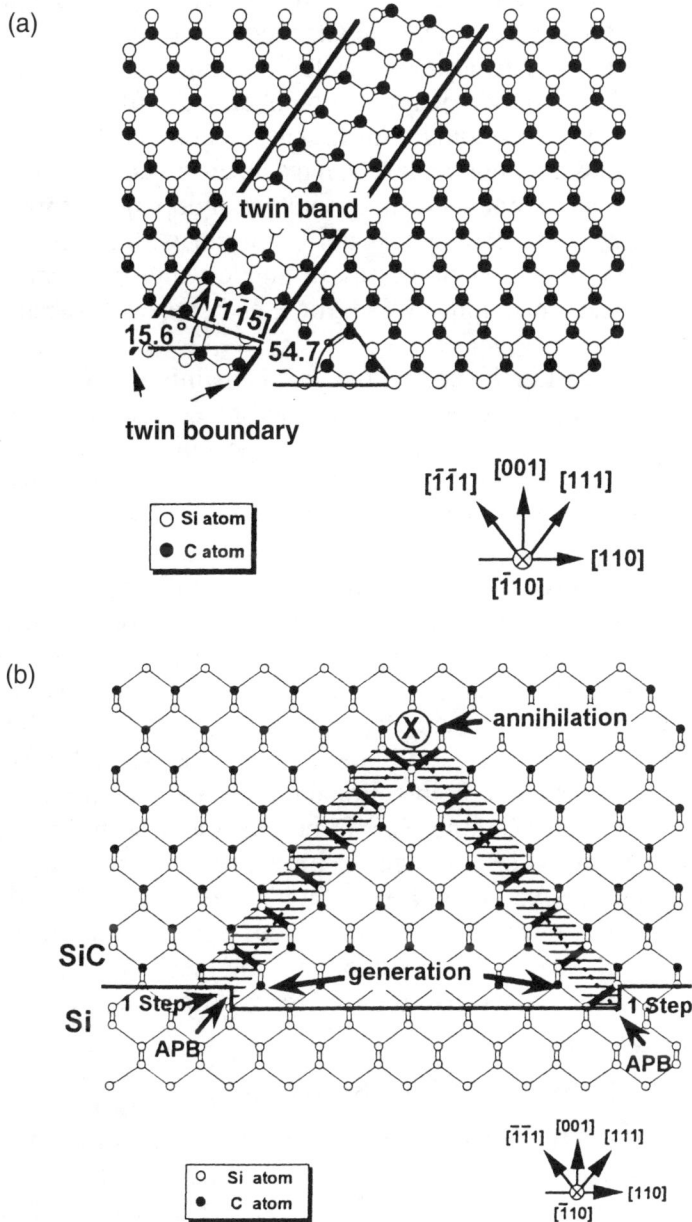

FIGURE 2.26   Illustration of crystal defects in 3C SiC: (a) twin boundaries; (b) anti-phase boundaries. From [30].

The doping capability of the VPE process has been shown to be sufficient for most device demands, as p- and n-type doping can be controlled over wide ranges, and high-purity material can be obtained. However, the lowest doping levels are not reachable after growth of intentionally doped material without cleaning of the reactor, as the dopants stick to the reactor walls and re-evaporate during the following runs. This limits the background doping to around $5 \times 10^{15}\,cm^{-3}$, as highly doped material is grown. Nevertheless, a doping range of almost five orders of magnitude over 30 nm can be obtained [26].

The uniformity of doping and thickness is crucial for device production, and acceptable levels have been reached. For the multi-wafer reactors with substrate rotation, a doping uniformity of within $\pm15\%$ and a thickness uniformity of within $\pm5\%$ are obtained. For optimized processes in single-wafer reactors the uniformity can reach similar values, or even better with substrate rotation.

The defect level in the epitaxial layers is mainly limited by the substrate quality, regarding both the micropipes and the polishing damage. Cleaning of the wafer, especially by removing particles before loading is essential. However, the VPE process can also enhance the existing defects or add new defects, especially if non-optimized growth parameters are used, or if reactor deposits are allowed to settle on the wafer as particles.

3C SiC grown on silicon substrates might be used for electronic devices, but the possible applications are limited due to the high concentration of stacking faults, twins and dislocations in the material. The intrinsic doping is n-type, at a level above $1 \times 10^{18}\,cm^{-3}$, mainly due to the lattice defects, and p-type doping can be obtained by intentional doping with aluminium or boron. Diodes and simple Si/SiC heterostructures have been reported, but efforts to make, for example, rectifying contacts to the 3C material have failed due to the high defect concentration [29,30]. Instead, micro-mechanical devices have been fabricated, where the etch selectivity to silicon and the mechanical characteristics are used, rather than the electronic properties.

> *Uniformity defined as: (maximum – minimum)/ mean with 5 mm edge excluded*

2.3.4 Comparison of techniques

Most important for epitaxial growth of SiC device structures is the possibility to control and reproduce key parameters, such as thickness, uniformity, doping and defects. The three commercial reactor types are compared with respect to these parameters in TABLE 2.6, and the running costs and potential for scaling up are also treated.

The growth rate and maximum thickness of the grown layers are to some extent coupled, as too low a growth rate will make thick layers less attractive to grow. On the other hand, the maximum thickness can be limited by defects formed during growth, e.g. by particle formation. The hot-wall reactor has been found to produce up to $50\,\mu m$ thick and defect-free layers, at growth rates up to $8\,\mu m/h$. The planetary reactor has shown similar

TABLE 2.6   Comparison of different reactor concepts for epitaxial growth of SiC device structures.

Parameter \ Technique	Hot wall	Planetary	Vertical
Growth rate	+ +	+ +	+
Maximum thickness	+ + +	+ +	+
Thickness uniformity	+ +	+ +	+ +
Background doping	+ +	+ +	+
n-type doping	+ +	+ +	+ +
p-type doping	−	−	+
Defect control	+ +	+	+ +
Process stability	+	+	+
Susceptor durability	−	−	+
Precursor consumption	+	+	− −
Equipment cost	+	−	−
Up-scaling potential	− −	+ +	+ +

results, but this requires an accurate control of the ceiling temperature to limit particle formation. The vertical reactor can be used for growth rates up to $5\,\mu$m/h, and the thickest layers presented are about $10\,\mu$m. The substrate rotation introduced in the planetary and vertical reactors gives thickness uniformities of better than $\pm5\%$ over six or seven wafers, while the hot-wall reactor handles single wafers with a similar uniformity, unless substrate rotation is introduced.

The background doping level has been extensively studied in the hot-wall reactor, and n-type levels in the low $10^{14}\,\mathrm{cm}^{-3}$ range are routinely obtained [19]. The planetary reactor has shown, at best, an n-type background doping in the same range, with an electron mobility of $950\,\mathrm{cm}^2/\mathrm{V}\,\mathrm{s}$ and a minority carrier lifetime of $3\,\mu$s for 4H at room temperature [20]. In the vertical reactor, the background doping is p-type, probably due to impurities from the heated susceptor, with typical levels around $1\times10^{15}\,\mathrm{cm}^{-3}$ [23].

Intentional n-type doping by nitrogen incorporation has been studied in all the reactors, and it has been shown to be controllable between about 1×10^{15} and $1\times10^{19}\,\mathrm{cm}^{-3}$, by independently changing the precursor flow and the C:Si ratio.

The nitrogen doping uniformity is within $\pm 10\%$ for all the wafers in the vertical and planetary reactors, and a similar uniformity is obtained over a single wafer in the hot-wall reactor. The p-type doping has been studied to much less an extent, and only limited information is available. A uniformity of within $\pm 15\%$ has been reported for the vertical reactor. The main problem is the background level (or memory effect) after the growth of a p-doped layer. This has been investigated for the vertical reactor, and a sharp turn-off from 1×10^{19} to $2 \times 10^{16} \mathrm{cm}^{-3}$ has been shown [23]. For the other reactors, with larger areas of heated graphite, the turn-off is expected to be slower and the background level higher.

Defect control has been considerably improved by introducing off-cut substrates, and by optimizing the cleaning procedure before growth, both ex-situ before loading, and in situ as a part of the growth sequence. In addition to this, the reactor growth parameters must be adjusted. From this point of view the reactors produce similar results. However, the planetary reactor will need more cleaning of deposits on the reactor walls. If not removed, the deposits form particles, which settle on the substrate and give rise to defects. This problem is minor in the vertical reactor, where no growth occurs on the cold reactor walls, and also in the hot-wall reactor, where deposits grow compact due to the high wall temperatures.

The process stability and reproducibility relates to stochastic variations in the temperature, the gas flows, etc., due to unintentional changes in the reactor geometry or in the gas-flow control system. Limited reports on variations in doping and thickness have been presented, and they indicate that the run-to-run variation is comparable to the uniformity over a wafer for all the reactors. There are also long-term variations in, for example, background doping due to the aging of the susceptor and other heated parts in the reactor. This has to be considered in the regular maintenance procedures, and influences the running cost of the reactor. The hot-wall reactor, which has a large graphite susceptor, and the planetary reactor with complex graphite parts, both suffer from graphite aging, while the molybdenum susceptor used in the vertical reactor has a much longer lifetime, and the graphite platter is comparatively inexpensive to replace.

The precursor and carrier gas consumption is five to ten times higher in the vertical reactor than in the horizontal reactors to obtain a similar uniformity, due to the high gas flows needed to overcome convection in the large reactor volume. This also puts higher demands on the pumps and filters connected to the reactor. This is inherent in the design, and can be comparable to the cost of graphite parts in the other reactors.

The reactor complexity mainly shows as a cost of the equipment. Hence the multi-wafer reactors are somewhat more expensive than the hot-wall reactor. However, the systems for gas delivery, reactor heating and process control are similar. In summary, this makes the hot-wall reactor a more economical choice if only single-wafer growth is considered.

In addition to the process performance, the possibility to scale up the reactor for production is also important for the future of the technology. Hence, both of the multi-wafer reactors have already shown the way, and the possibility to further increase the number or size of the substrates is clearly shown for the III–V technology, on which the SiC epi reactors are based.

2.3.5 Overcoming problems

2.3.5.1 Reactor design

A fundamental advantage with the vertical and the hot-wall reactors over the planetary reactor is the smaller number of defects due to ceiling deposits, which easily peel off and form particles. An improvement to the planetary reactor design is the introduction of a separate ceiling heater, which could be temperature controlled independently of the substrate heater. This makes the process conditions more similar to those of the hot-wall reactor, but with a considerably improved control of the temperature boundary conditions.

For the hot-wall reactor, the main limitation is the thickness and doping uniformity, which can be considerably improved by introducing substrate rotation. Rotation will efficiently average the non-uniformity in the gas-phase depletion and temperature fields. Using a planetary reactor with separate ceiling heater can solve the other drawback with the hot-wall reactor, the limited possibility to scale up the process.

The vertical reactor has limitations in maximum growth rate, which is related to the substrate temperature. Due to the cold walls, the substrate surface maintains a considerably lower temperature than the susceptor. By increasing the susceptor temperature, and the precursor supersaturation accordingly, the growth rate could probably be increased, but with an increase in the impurity level, due to enhanced evaporation from the molybdenum susceptor. To overcome this, the susceptor material has to be chosen with care, and the process optimized.

For all reactors, a major issue is the use of hydrogen gas at a temperature where the molecules decompose into atomic hydrogen, which is highly reactive and attacks the heated graphite and SiC-coated reactor parts. This limits the reactor lifetime, and both hydrocarbons and impurities (mainly boron, aluminium and titanium) are released from the graphite. These will be present at the growing surface, and can affect the crystal quality.

Different ways to circumvent those problems have been suggested. Solid refractory metal susceptors have been used, as well as improved graphite coatings including SiC or metal carbides. A free-standing polycrystalline SiC plate, about 800 μm thick, is commercially available and can be placed on top of a coated or non-coated graphite susceptor to protect the graphite or coated graphite surface from erosion.

A drawback with the SiC coating, and to a lesser extent also with the polycrystalline SiC plate, is the comparatively short lifetime. The material readily sublimes, and deposits on the backside of the somewhat colder substrate, forming a polycrystalline SiC layer. If the backside of the substrate is to be used for contacts, or a planar surface is required, this layer must be removed by grinding or polishing. Despite the problems, these solutions are the most commonly used.

Susceptors or reactor parts made of refractory metals, e.g. molybdenum or tantalum, are used in some applications, but the impurity levels are comparable to those obtained from uncoated graphite. A combination, where a molybdenum platen is used together with coated graphite parts, is used in the vertical reactor. Here, the substrate is not exposed to the metal, and the construction allows the use of inexpensive coated graphite parts [23]. However, the design has to be improved in order to reduce the background doping, e.g. by applying a high-purity free-standing SiC platter on the susceptor.

More promising, but less evaluated are metal carbides, such as tantalum carbide, TaC, and niobium carbide, NbC. These coatings are found to possess lifetimes more than ten times longer than for SiC coatings, and material grown on these susceptors shows a smooth morphology and considerably lower unintentional doping levels than material grown using other susceptors [31].

The gas-foil rotation mechanism used in the planetary reactor works by introducing a carrier gas through narrow channels drilled in the graphite. These channels could not, for practical reasons, be coated by any material, and a way to avoid early erosion is to use an inert gas, e.g. argon or helium, for the gas-foil rotation.

2.3.5.2 *Doping control*

Doping control is intricate in the SiC process. The background doping can be limited by using purified gases, and high-grade materials in the critical parts of the reactor. More problematic can be memory effects from earlier growth steps where dopants have been intentionally introduced.

By using site-competition epitaxy, the impurity level can be controlled by adjusting the C : Si ratio in the gas phase. Incorporation of the p-type dopants aluminium and boron, is increased at a high C : Si ratio, while the n-type dopant nitrogen is increased at a low C : Si ratio. Hence, the C : Si ratio should be chosen to limit the dominating dopant to grow low-doped material, while intentionally doped material should rather be grown under the C : Si ratio most suited for the dopant of choice.

A memory effect occurs when dopants are adsorbed on the reactor walls, and later re-evaporate to be incorporated in the growing crystal. This effect is less pronounced for nitrogen than for aluminium and boron. It can be reduced by etching the reactor after the doped layer has been grown, either by a hydrogen or by a hydrochloric acid etch. A combination of an etch and an active C : Si ratio control has been found to allow very abrupt doping transients [26]. However, this method is not sufficient if a background doping of below $2 \times 10^{15}\,cm^{-3}$ is required. In this case, a reactor cell dedicated for low background doping could be the solution. A two-reactor cell option is usually suggested for the hot-wall reactor, where the memory effect of p-type dopants can be severe.

To reduce the nitrogen background, a load lock or efficient purging of the reactor to reduce the atmospheric contamination before growth is the preferred solution.

2.3.5.3 *Defect reduction*

A low defect concentration relies on good substrate handling. The substrate surface should be clean and free from polishing damage, scratches, or impurities when the growth starts. Etching in hydrofluoric acid before loading the wafer into the reactor is common. It will remove the native oxide and some metal traces. In addition, an in situ etch before growth is usually required. It has been found that a pure hydrogen etch can leave silicon droplets, which act as nucleation centres for defects on the surface. The addition of hydrochloric acid or propane eliminates this problem [20].

The use of off-cut substrates with a clean surface and a low micropipe density is a prerequisite for growing defect-free crystals. In addition, the growth parameters must be chosen with care, by employing a good balance between the growth temperature and the supersaturation, i.e. the total vapour pressure of the precursors and a well-balanced C : Si ratio.

2.4 RECENT DEVELOPMENTS AND FUTURE TRENDS

In crystal growth of SiC, regardless of growth technique, improvement of the results relies heavily on the knowledge and understanding of the process. The knowledge may be obtained by materials characterisation; equally important will be in situ characterization of the growth process. An obvious first step is to make a detailed mapping of the temperature in the growth cell, and on the substrate. More information will be given by chemical analysis and surface analysis during growth. Some investigations have been made, but much more remains to be done, especially characterization of the chemical reaction pathways and surface reactions during growth.

A detailed understanding of the growth also needs theoretical modelling of the process. Hence, the thermodynamic, kinetic and gas-flow dynamic aspects must be taken into account to establish the model. The outcome can include growth rate and impurity incorporation, polytype stability and formation of defects, e.g. how the origin of silicon droplets and graphite precipitates form micropipes and dislocations, and thermoelastic stress in the grown crystal. All of these are related to the growth parameters.

The acquired understanding can be used for further development of the growth equipment and for tuning of the process parameters, with the goal of growing defect-free material, of large diameter and of excellent uniformity and homogeneity.

A more spectacular future development would be the possibility to independently control the polytype grown, by controlling one or a few critical growth parameters. This would enable growth of heterostructures, where the lattice mismatch is perfectly zero, and a bandgap offset of up to 1 eV (which is the bandgap difference between 3C and 4H SiC) could be obtained. Single layers of 3C on on-axis 6H and 4H SiC are easily grown, and also 4H and 6H have been grown on 3C, which indicates that the ideal heterostructure could be within reach [32].

2.5 CONCLUSIONS

A prerequisite to realising the full potential of SiC electronic applications is the availability of large-diameter defect-free substrates, with well-controlled doping concentrations. On these substrates it should be possible to grow epitaxial layers with highly uniform thicknesses and doping over large areas. The doping incorporation profiles should be possible to custom design, and the background impurity level should be below the level determined by the most demanding device. The possibility to obtain material with these ideal parameters strictly depends on the maturity of the technology developed for crystal growth.

The dominating bulk growth technique, the seeded sublimation growth, has recently improved considerably, and can produce a substrate material of a quality approaching that demanded. However, the substrate diameter should still be increased, and the defect concentration decreased. Micropipes will probably soon become an outdated problem, while the less-apparent dislocations will be a focus for development for many years to come.

The alternative techniques, liquid phase epitaxy, and high-temperature chemical vapour deposition, both have great potential. Currently their main importance is for specific applications, such as high-purity material grown by HTCVD and low micropipe density material grown by LPE. However, it is most likely that these techniques will develop and become valuable complements to the dominant seeded sublimation growth.

For epitaxial growth, vapour phase epitaxy is the only considered technique. A few reactor concepts have been developed. They present different solutions to the problem of handling the highly reactive hydrogen at the growth temperature, and still produce a high-purity and defect-free material. In the hot-wall reactor and the planetary reactor with active ceiling heating, the reactions take place in an efficiently heated cavity containing the substrate. A high-quality material can be grown at high growth rates, but the large heated graphite areas will enhance the wall deposits and contribute to doping memory effects. An alternative is the vertical reactor, which reduces the heated areas and provides a cold-wall cavity with a high gas flow. Hence, a good handling of doping transients can be obtained, but at the cost of a large gas consumption.

For the future, the development of the growth technologies relies on an increased process understanding, both from in situ characterisation and from theoretical modelling. The outcome will be larger wafer areas, improved uniformity, reduced defect concentration and superior doping control.

REFERENCES

[1] W.F. Knippenberg [*Philips Res. Rep. (Netherlands)* vol.18 (1963) p.161–274]
[2] Yu.M. Tairov, V.F. Tsvetkov [*J. Cryst. Growth (Netherlands)* vol.52 (1981) p.146–50]
[3] S.Yu. Karpov, Yu.N. Makarov, M.S. Ramm [*Phys. Status Solidi B (Germany)* vol.202 (1997) p.201–20]

[4] A.S. Segal et al [*Mater. Sci. Eng. (Netherlands)* vol.B61–62 (1999) p.40–3]

[5] R. Yakimova et al [*Mater. Sci. Eng. (Netherlands)* vol. B61–62 (1999) p.54–7]

[6] Yu. Vodakov, A.D. Roenkov, M.G. Ramm, E.N. Mokhov, Yu.N. Markov [*Phys. Status Solidi B (Germany)* vol.202 (1997) p.177–200]

[7] G. Augustine, H.McD. Hobgood, V. Balakrishna, G. Dunne, R.H. Hopkins [*Phys. Status Solidi B (Germany)* vol.202 (1997) p.137–48]

[8] R.A. Stein, P. Laning [*J. Cryst. Growth (Netherlands)* vol.131 (1993) p.71–4]

[9] M. Kanaya, J. Takahashi, Y. Fujiwara, A. Moritani [*Appl. Phys. Lett. (USA)* vol.58 (1991) p.56–8]

[10] M. Dudley et al [*Appl. Phys. Lett. (USA)* vol.75 (1999) p.784–6]

[11] D.H. Hoffmann, M.H. Müller [*Mater. Sci. Eng. (Netherlands)* vol.B61–62 (1999) p.29–39]

[12] A. Ellison et al [*Mater. Sci. Eng. (Netherlands)* vol.B61–62 (1999) p.113–20]

[13] J. Takahashi, N. Ohtani [*Phys. Status Solidi B (Germany)* vol.202 (1997) p.163–75]

[14] A. Ellison [Ph.D. Thesis, Linköping Studies in Science and Technology Dissertation no. 599, Linköping University, 1999]

[15] L. Zhou, V. Audurier, P. Pirouz, A. Powell [*J. Electrochem. Soc. (USA)* vol.144 (1997) p.L161–3]

[16] T. Kimoto, A. Itoh, H. Matsunami [*Phys. Status Solidi B (Germany)* vol.202 (1997) p.247–62]

[17] N. Nordell, A. Schöner, S.G. Andersson [*J. Electrochem. Soc. (USA)* vol.143 (1996) p.2910–9]

[18] Epigress AB (member of the Aixtron group), SE – 223 70 Lund, Sweden

[19] O. Kordina et al [*Phys. Status Solidi B (Germany)* vol.202 (1997) p.321–34]

[20] A.A. Burk Jr., L.B. Rowland [*Phys. Status Solidi B (Germany)* vol.202 (1997) p.263–79]

[21] Emcore Corp, 394 Elizabeth Avenue, Somerset, NJ 08873, USA

[22] R. Rupp, Yu.N. Makarov, H. Behner, A. Wiedenhofer [*Phys. Status Solidi B (Germany)* vol.202 (1997) p.281–304]

[23] S. Karlsson, N. Nordell, F. Spadafora, M. Linnarsson [*Mater Sci. Eng. (Netherlands)* vol.B61–62 (1999) p.143–6]

[24] M.D. Allendorf, R.J. Lee [*J. Electrochem. Soc. (USA)* vol.138 (1991) p.841–52]

[25] M.I. Chaudhry, R.J. McCluskey, R.L. Wright [*J. Cryst. Growth (Netherlands)* vol.113 (1991) p.120–6]

[26] N. Nordell, A. Schöner, M.K Linnarsson [*J. Electron. Mater. (USA)* vol.26 (1997) p.187–92]

[27] D.J. Larkin [*Phys. Status Solidi B (Germany)* vol.202 (1997) p.305–20]

[28] A. Schöner, K. Rottner, N. Nordell, M. Linnarsson, Ch. Peppermüller, R. Helbig [*Diam. Relat. Mater. (Netherlands)* vol.6 (1997) p.1293–6]

[29] S. Nishino, H. Suhara, H. Ono, H. Matsunami [*J. Appl. Phys. (USA)* vol.61 (1987) p.4889–93]

[30] H. Nagasawa, K. Yagi [*Phys. Status Solidi B (Germany)* vol.202 (1997) p.335–58]

[31] B.E. Landini [*J. Electron. Mater. (USA)* vol.29 (2000) p.384–90]

[32] A. Fissel, B. Schröter, U. Kaiser, W. Richter [*Appl. Phys. Lett. (USA)* vol.77 (2000) p.2418–20]

Chapter 3

Ion implantation and diffusion in SiC

A. Schöner

3.1 CHAPTER SCOPE

In conventional semiconductors like silicon and gallium arsenide, the processes for ion implantation and diffusion are well established. Because of the inherent material properties of SiC, these processes cannot be transferred. In this chapter the processes for ion implantation and diffusion of SiC will be described. The main emphasis is on the outcome of the processes rather than the technological realisation. The idea is to provide a basic understanding of the doping process, which is of major concern for the design of electronic devices.

Ion implantation has the advantage over other doping processes that basically all stable elements of the periodic table can be implanted. In addition, lateral structuring and doping of selected areas is possible through masking techniques. Due to low diffusion coefficients of the main dopants and the high temperatures in excess of 2000 °C needed, diffusion is not a suitable process for doping of SiC. However, diffusion mechanisms have to be investigated as small impurities like hydrogen, lithium, beryllium, and boron can move during a high-temperature treatment. High-temperature annealing is applied after ion implantation to activate the dopants and remove the irradiation induced damage. Hence, the main focus of this chapter is doping by ion implantation.

To tailor the electrical properties of device areas, the knowledge of lateral and depth distribution as well as the electrical properties of impurities and defects are important issues, which have to be studied carefully. A short description of selected characterisation methods and the difficulties in interpreting the results is included in the chapter.

In Section 3.2 on ion implantation the electrical properties of the main shallow dopants governing the conductivity type as well as of deep extrinsic and intrinsic impurities will be reviewed.

Extrinsic deep centres are interesting to tailor the resistivity of certain device regions and intrinsic defects are part of the ion implantation technology. Intrinsic defects are formed by residual damage of the crystal structure caused by the ion bombardment.

3.2 ION IMPLANTATION

In semiconductor processing ion implantation is the key technology for doping, besides diffusion and crystal growth. Doping concentrations and doping profiles can be adjusted reproducibly and varied over a wide range. The advantage of ion implantation is the possibility to dope selected areas through masking techniques, avoiding lateral structuring with wet and dry etching techniques. The masking can be done by either deposited metals with a high mass (e.g. gold) or thermally grown or deposited silicon dioxide. The masking films must be thicker than the maximum travelling distance of the implanted ions in the film and depend on the applied acceleration energies. The mask should not consist of element species that are dopants or recombination centres in SiC. The patterning of the mask is done by common lithography techniques.

3.2.1   Ion implantation of SiC: general process

The technique of shooting ions into SiC crystals is not different from the one commonly used in other semiconductor processing. The difference is in the process for dopant activation and re-crystallisation.

After the irradiation with ions, the implanted atoms occupy predominantly interstitial lattice sites. These interstitial atoms do not affect the electrical material properties as long as they are not part of more complex defects. In addition, irradiating the material with ions causes damage to the crystal structure, depending on the ion mass, ion energy, and the implanted fluence. Silicon and carbon atoms are kicked out from their lattice positions. The kicked-out silicon and carbon atoms reside to a large extent on interstitial lattice sites, leaving vacant lattice places behind.

To activate implanted atoms and to reduce the induced damage, the ion-implantation process combines the introduction of the doping species with a thermal annealing step. The thermal annealing allows the re-crystallisation of the material. During the re-crystallisation, the interstitial doping atoms compete with interstitial silicon and carbon for free lattice places (recombination with vacancies). The atoms ending up on lattice places become electrically active. Impurities consisting of isolated dopant atoms (see Sections 3.2.2.2, 3.2.2.3, 3.2.2.4 and 3.2.3.3), intrinsic defect centres (see Section 3.2.3.4), and complexes of both are formed.

In SiC, temperatures as high as 2000 °C are employed for the thermal post-implantation anneal under argon at atmospheric pressure. Processing equipment in standard semiconductor technology is typically designed for temperatures up to 1300 °C, so conventional annealing equipment cannot be applied to SiC. In addition, the melting point of SiC is around 3000 °C and a pressure of 30 bar [1]. At atmospheric pressure and temperatures above 1300 °C, SiC sublimes directly from the solid phase into the gas phase. It was observed that the SiC surface degrades under thermal treatment at temperatures above 1300 °C in inert gas atmosphere (e.g. argon), as well as in vacuum. Silicon seems to evaporate and a carbon layer is left behind on the surface [2]. This carbon layer is difficult to remove by wet etching techniques. As is known from III–V compound semiconductors, keeping an overpressure of the evaporating species in the furnace atmosphere prevents the surface from degrading. A silicon overpressure can be accomplished by either adding silicon-containing gases (e.g. silane SiH_4) to the system or using containers made from polycrystalline SiC. Depositing a sacrificial layer on the implanted SiC, which is stable at high temperatures and can be easily removed after the anneal, is another method of avoiding the evaporation of silicon [3].

Comparing different techniques for the post-implantation anneal, it seems that better results with respect to surface degradation and morphology can be obtained, when the anneal is done under gas-flow conditions and not in a static atmosphere. This is probably accomplished by a slight etching of the surface. Hence, suitable process systems for thermal annealing of implanted SiC can be designed similar to epitaxial growth reactors. Such reactors can typically be used for annealing of SiC up to temperatures of 1800 °C. For higher temperatures, equipment similar to sublimation systems used for bulk growth of SiC can be employed.

Besides the high-temperature post-implantation anneal, a considerable further improvement in electrical activation of dopants and the quality of the re-crystallisation of the material was achieved by implanting the ions at elevated temperatures [4]. The material is heated to temperatures in the range of 500 °C to 1000 °C giving the crystal the possibility to instantly heal itself to some extent. The material temperature is chosen with respect to the doping species to implant. Implantation at elevated temperatures imposes constraints on the material used for masking, since the masking material has to withstand the applied temperatures and should not lose its stopping properties.

Implantation at temperatures above 500 °C together with the post-implantation anneal treatment at temperatures of around 1700 °C for 10 min under argon and silane flow is currently one of the processes that gives the best results in terms of surface morphology, doping activation and damage reduction.

3.2.2 Shallow impurities for n-type and p-type conductivity

To achieve n-type and p-type conductivity, the semiconductor material has to be doped with donors and acceptors, respectively. Acceptors are elements that have at least one electron less in the outer shell than the semiconductor atoms, donors having one electron more. As SiC consists of the group IV elements silicon and carbon having four electrons in the outer shell, the group III elements boron, aluminium, gallium and indium act as acceptors. Donors are formed by the group V elements nitrogen, phosphorus, arsenic and antimony. Aluminium and boron are the acceptor species that have been investigated in detail and are the subject of Section 3.2.2.2. The donor impurities nitrogen and phosphorus will be discussed in Section 3.2.2.3. Aluminium, boron, nitrogen and phosphorus are the shallow impurities, which are mostly used in doping processes for device fabrication.

3.2.2.1 *Characterisation of shallow impurities and implanted layers*

The design of devices implies the knowledge of the shape of the doping profile, the doping concentration, and the electrical activation of dopants. The basic principle of a selection of suitable methods to characterise the material with respect to the device design parameters and the difficulties of applying these methods to SiC is the content of the following pages.

Secondary ion mass spectrometry (SIMS)

The atomic concentration and the in-depth distribution of impurities can be detected by secondary ion mass spectrometry (SIMS). The SIMS equipment is a sensitive mass spectrometer to analyse atoms sputtered from the material using a primary ion beam. A schematic picture of a SIMS instrument is shown in FIGURE 3.1. Small craters of the order of 100 μm edge length are sputtered into the semiconductor material by using typically an oxygen or caesium ion beam. The sputtered atoms are ionised and collected as secondary ions in a mass spectrometer, filtering out the mass to investigate. The detection of the filtered ions is done in the case of low impurity concentrations with an

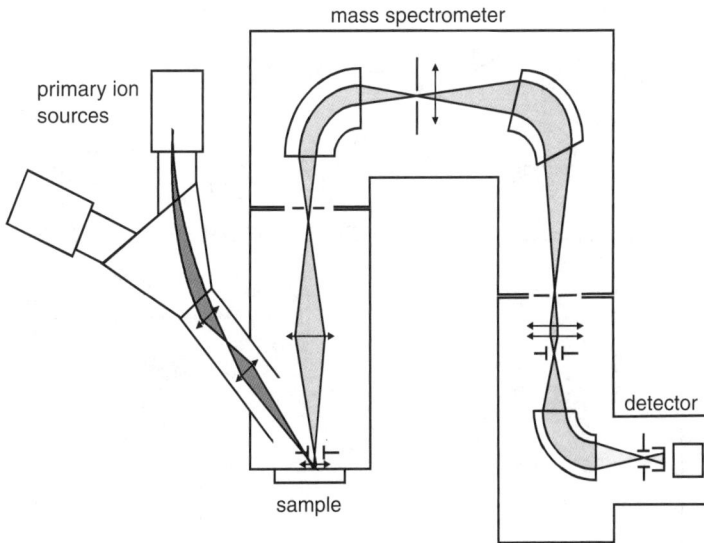

FIGURE 3.1 Schematic of a SIMS instrument.

electron multiplier or for high impurity concentrations with a Faraday cup.

Assuming a constant sputter rate and measuring the crater depth with a stylus profilometer, the depth distribution of the dopants is determined. The correlation of the counted ions with the atomic concentration in the semiconductor material is done by implanted calibration standards.

Hall effect measurements

For the characterisation of the electrical properties active doping concentration, dopant ionisation energy, conductivity, free charge carrier concentration, and charge carrier mobility, Hall effect measurements can be applied. Hall effect measurements exist in various methods, which differ mainly by the geometry of the used structures. All structures have in common that at least four electrical contacts with ohmic current–voltage characteristic are required. Van der Pauw structures (FIGURE 3.2) and bar-type structures are the most commonly used Hall effect structures. Hall effect with the van der Pauw geometry measures directly the material resistivity, the conductivity type, the free charge carrier concentration, and the charge carrier mobility. In addition, it gives information about the lateral conductivity uniformity in the material. Bar-type Hall effect structures can be oriented in crystallographic directions. The material resistivity, the free charge carrier concentration and the carrier mobility in these directions is determined. The requirements for both methods are a homogeneous thickness and doping concentration.

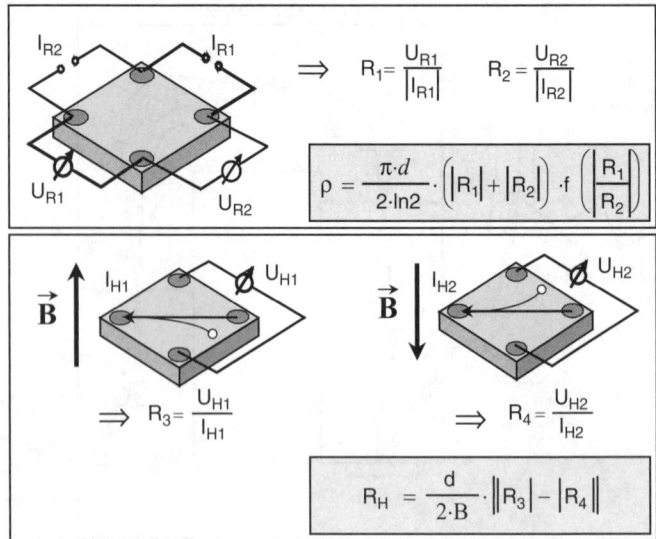

FIGURE 3.2   Principle of van der Pauw Hall effect measurements of a squared sample with homogeneous thickness d. ρ is the measured resistivity of the material and R_H the Hall coefficient, which is correlated to the free carrier concentration.

Charge neutrality equation
for n-type semiconductors assuming an intrinsic carrier concentration much smaller than the doping concentration:

$$n + N_{CA}$$

$$= \sum_i \frac{N_{Di}}{1 + \dfrac{g_{Di}\, n}{N_C} \exp\!\left(\dfrac{\Delta E_{Di}}{k_B T}\right)};$$

$$N_C = 2 M_C \left(\frac{2\pi m_{ec} k_B T}{h^2}\right)^{3/2}$$

n	free electron concentration
N_{CA}	concentration of compensating acceptors
N_{Di}	concentration of the i-th donor
ΔE_{Di}	ionisation energy of the i-th donor
g_{Di}	degeneracy factor for the i-th donor
N_C	conduction band effective density of states
m_{ec}	effective electron mass
M_C	number of equivalent conduction band minima
k_B	Boltzmann constant
T	temperature
h	Planck constant

The doping and thickness homogeneity is detected by the van der Pauw method in conductivity variations of the two perpendicularly oriented directions of the supplied current (I_{R1} and I_{R2} in FIGURE 3.2). The shape of the sample is not significant for the van der Pauw Hall effect, but in the case of bar-type structures, one edge dimension has to be at least a factor of 5 longer than the other two, if the electrical properties are measured versus the crystallographic direction.

To get information about the shallow dopants in the sample, temperature-dependent Hall effect measurements are performed. From temperature scans, the doping concentration and ionisation energy can be derived by a fit of the charge neutrality equation to the experimental free charge carrier concentration data. To perform this fit, the material parameters effective density of states in the transport band (valence or conduction band) and Hall scattering factor have to be known. The effective density of states includes the effective charge carrier mass and, in the case of the conduction band, the number of equivalent conduction band minima in the Brillouin zone. The effective charge carrier masses for the most common SiC polytypes are available from theoretical band structure calculations [5].

The remaining unknown parameter is the Hall scattering factor. The Hall scattering factor is the ratio of the mobility determined by Hall effect to the carrier drift mobility and depends on the magnetic field, the doping concentration, and the temperature.

Presently, the Hall scattering factor is only determined for low-doped n-type epitaxial SiC-layers [6]. No values are available for p-type and high n-type doped SiC material. When the Hall scattering factor is unknown, it is set equal to 1.

As mentioned, the requirements for standard Hall effect measurements are that the material under investigation is homogeneously doped and the thickness is known. Hence, thin epitaxially grown or implanted layers have to be insulated from the substrate. This insulation could be done by using semi-insulating wafers or substrates with opposite conductivity type to the layer under investigation. In the latter case, pn-junctions are used to prevent the applied current from penetrating into the substrate. For implanted SiC layers, the insulation by a pn-junction is not always sufficient. The reason for this is that the resistivity of implanted SiC layers (especially for p-type layers) is very high and of the order of the pn-junction resistance, due to the small thickness and the deep acceptor levels (Section 3.2.2.2). In addition, the pn-junctions are leaky due to remaining damage from the implantation. In the case of insufficient insulation of the implanted layer from the substrate, a two-layer analysis of the Hall effect data must be used to calculate the real electrical properties of the layer.

Capacitance–voltage (CV) measurements

The net doping concentration in the space charge region of Schottky contacts or pn-junctions is a result from capacitance-voltage (CV) measurements. The net doping concentration is the difference of the electrically active concentration of donors and acceptors. In the case of n-type material, the concentration of donors outbalances the concentration of acceptors, and vice versa for p-type material.

CV measurements can be applied up to net doping concentrations of $5 \times 10^{18}\,\text{cm}^{-3}$. By increasing the reverse bias voltage on the Schottky contact or pn-junction, the depletion region is increased and the in-depth distribution of the net doping is determined.

CV measurements on implanted samples require a good annealing of the induced damage. If the remaining damage is high, the prepared Schottky contacts and pn-junctions are leaky. A significant leakage current disturbs the capacitance measurement and the consequence is a false estimation of the net doping concentration. In addition, the induced damage can form electrically active deep centres, which act as compensation. If the compensation is negligible, the net doping concentration obtained by CV is equal to the doping concentration determined from the fit

---

Charge neutrality equation
for p-type semiconductors assuming an intrinsic carrier concentration much smaller than the doping concentration:

$$p + N_{CD}$$

$$= \sum_i \frac{N_{Ai}}{1 + \dfrac{g_{Ai}p}{N_V} \exp\left(\dfrac{\Delta E_{Ai}}{k_B T}\right)};$$

$$N_V = 2\left(\frac{2\pi m_{ep} k_B T}{h^2}\right)^{3/2}$$

p	free hole concentration
N_{CD}	concentration of compensating donors
N_{Ai}	concentration of the i-th acceptor
ΔE_{Ai}	ionisation energy of the i-th acceptor
g_{Ai}	degeneracy factor for the i-th acceptor
N_V	valence band effective density of states
m_{ep}	effective hole mass
k_B	Boltzmann constant
T	temperature
h	Planck constant

of the neutrality equation to the free charge carrier concentration measured by Hall effect.

Measurement of the dopant activation

If all atoms introduced by the doping process reside on electrically active lattice sites, the atomic concentration determined by SIMS and the doping concentration obtained by Hall effect should be equal. This is not necessarily always the case. As mentioned before, after ion implantation the introduced atoms occupy predominantly interstitial lattice sites. During the high-temperature anneal they compete with interstitial silicon and carbon atoms for empty lattice places. The number of dopant atoms that manage to reside on a lattice place divided by the total number of introduced dopant atoms is called the degree of activation, which is a key figure for the quality of the doping process.

To accurately determine the dopant activation after ion implantation and subsequent high-temperature anneal at least two kinds of measurements should be performed on the same set of samples. The first measurement determines the atomic doping concentration and the dopant distribution (for example with SIMS) and the second measurement the electrically active doping concentration (for example with Hall effect or CV).

Investigations on dopant activation have been done in past years extensively for several impurities in SiC. Unfortunately, the determined activation values are often reported to be higher than 1, which makes no sense physically. Although SIMS measurements have a relatively high uncertainty, the main systematic error is probably in the Hall effect measurements with the uncertainty in material parameters (e.g. Hall scattering factor) and the difficulties in completely insulating thin high-resistive layers from the substrate.

3.2.2.2 Acceptor dopants in SiC: aluminium and boron

The group III elements boron, aluminium, gallium and indium act as acceptors in SiC. Aluminium and boron are the acceptor species that have been investigated in detail and are mostly used for doping in device processing. The acceptor dopants aluminium and boron are the subject of this section. Reference [7] contains a review and summary of investigations on implantation doping of aluminium and boron in SiC.

Projected range of aluminium and boron

The projected range is the average distance a doping atom penetrates into the crystal during ion implantation. The peak maximum in single-energy implanted profiles is a good estimate for the projected range, if channelling effects can be neglected.

FIGURE 3.3 shows depth profiles obtained by SIMS of aluminium and boron implanted 4H-SiC epitaxial layers. The single-energy implants were conducted at 600 °C with energies of 200 keV, 500 keV, and 1 MeV and a fluence of $1 \times 10^{14} \, cm^{-2}$ and $1 \times 10^{15} \, cm^{-2}$, respectively. No post-implantation anneal was performed on these samples. Because boron is much lighter and smaller than aluminium, the projected range for boron is about a factor of 2 higher than the projected range for aluminium, when implanted with the same energy. To form implanted p-type SiC layers with a thickness of more than 1 μm, energies of around 1 MeV are needed in the case of aluminium and about 0.5 MeV in the case of boron. Hence, if thick p-type layers are required, boron has an advantage over aluminium, because it can be implanted deeper with accelerators available in semiconductor processing. In addition, boron causes less damage to the crystal due to its size and mass. But boron as a dopant has the disadvantage of forming the impurity level with the higher ionisation energy compared to aluminium.

Electrical properties of aluminium and boron

Devices consist usually of areas with different resistivity. The resistivity is largely governed by the concentration of the introduced dopants and their ionisation energy. Aluminium and boron

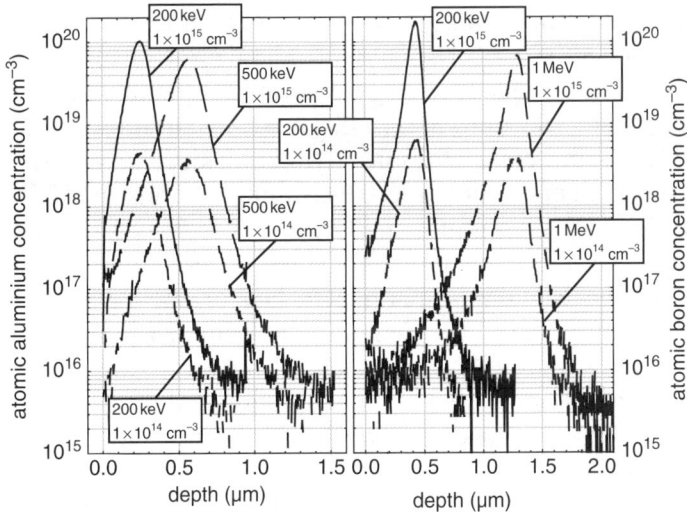

FIGURE 3.3   Depth profiles observed by secondary ion mass spectrometry (SIMS) from 4H-SiC epitaxial layers implanted with aluminium and boron. The parameters of the implantation are summarised in the figure.

TABLE 3.1 Ionisation energies of shallow acceptor impurities in the semiconductors 4H-SiC, 6H-SiC, silicon, and gallium arsenide [7–9].

Semi-conductor	Shallow acceptor species	Ionisation energy (meV)
4H-SiC	aluminium	200
	boron	285
	gallium	290
6H-SiC	aluminium	240
	boron	300
	gallium	300
Si	aluminium	67
	boron	45
	gallium	72
GaAs	carbon	26
	beryllium	28
	magnesium	28
	zinc	31

in SiC have much higher ionisation energies compared with shallow acceptors in the semiconductors silicon and GaAs (TABLE 3.1).

The valence band structures of the SiC polytypes are similar. The electrical properties of acceptor dopants therefore do not vary much with the polytype. In electrical measurements, aluminium is usually observed as a single acceptor level located about 200 meV to 240 meV above the valence band. Aluminium atoms occupy places in the crystal where normally a silicon atom is placed (silicon sublattice site). From atomic-size considerations, aluminium is too large and would lead to a high stress in the crystal when placed on a carbon site.

Boron forms two electrical levels in the lower half of the SiC bandgap. The shallower level is located around 300 meV above the valence band observed by Hall effect. The second level is 550–650 meV above the valence band observed with deep level transient spectroscopy (DLTS) (for a description of DLTS see Section 3.2.3.1). The microscopic structure of the shallower level is boron residing on a silicon site, where it acts as an acceptor. The microscopic structure of the deeper level is not resolved yet, but there is evidence in the literature that it is also of an acceptor nature. Boron placed on a silicon site with a neighbouring carbon vacancy or boron residing on a carbon site are possible models discussed in the literature. Recent theoretical calculations on binding energy and stability of boron-related complexes favour the model of boron on a carbon site [10].

Because of the large energy distance between the valence band maximum and the acceptor levels, only a small part of the introduced atoms is ionised at room temperature and contributes to the conductivity of the material. The degree of ionisation can be calculated from the charge neutrality equation for a single acceptor (see Section 3.2.2.1). The concentration of ionised acceptors is the concentration of free holes in the valence band plus the concentration of compensating donors. FIGURE 3.4 shows for both aluminium and boron the degree of ionisation as a function of the temperature for three different doping concentrations. The calculation parameters are given in FIGURE 3.4. At room temperature the degree of ionisation for aluminium is considerably higher than for boron. To achieve similar conductivity with boron to that with aluminium, almost an order of magnitude higher concentration has to be implanted. Although aluminium is a larger atom than boron and therefore causes more damage, the lower concentration to implant implies similar electrical properties of the implanted area with less damage to the crystal. In addition, as shown in Section 3.3.3, aluminium does not migrate in the SiC crystal. It stays where it is

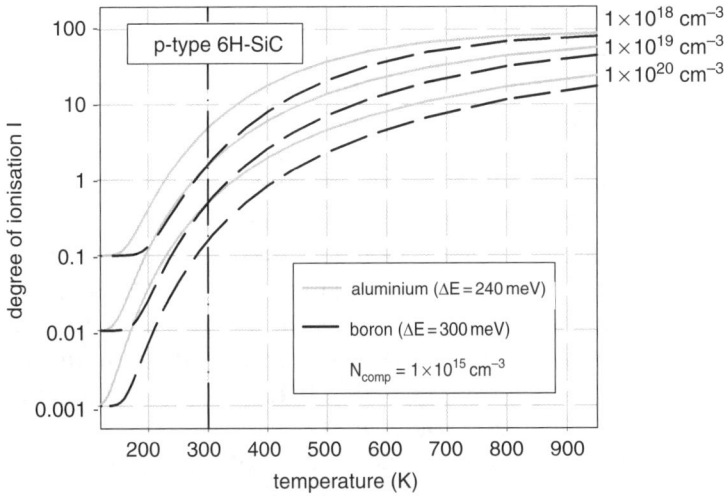

FIGURE 3.4   Calculated degree of ionisation I as a function of the temperature for aluminium- and boron-doped p-type SiC. The calculation parameters are summarised in the figure.

Degree of ionisation for one acceptor level ΔE_A of concentration N_A:

$$I = \frac{N_A^-}{N_A} = \frac{1}{1 + \frac{g_A p}{N_V} \exp\left(\frac{\Delta E_A}{k_B T}\right)}$$

I	degree of ionisation
p	free hole concentration
N_A^-	concentration of ionised acceptors
N_A	total concentration of acceptors
ΔE_A	ionisation energy of the acceptor
N_V	effective density of states of the valence band
g_A	degeneracy factor (if the split off band can be neglected g is set to 4 for acceptors)
k_B	Boltzmann constant
T	temperature

implanted. Therefore, if high-conductive p-type layers are required in the device design, the preferred acceptor dopant in SiC is aluminium.

Carbon and silicon co-implantation

As discussed above, boron forms two energy levels in the bandgap of SiC. To achieve a high p-type conductivity with boron as an acceptor dopant, it is beneficial when boron is incorporated preferentially at silicon lattice sites, forming the acceptor level with an ionisation energy of approximately 300 meV. Optimising the post-implantation anneal conditions is not enough to incorporate boron exclusively on silicon lattice sites. Still, a considerable amount of boron will form the deep D-centre.

From epitaxial growth it is known that the incorporation of impurities on a sublattice site is controlled by the ratio of carbon and silicon atoms (C/Si ratio) in the gas phase. The C/Si ratio is set by the precursor flow rate of silicon and carbon. This process is called site-competition epitaxy.

In ion implantation, the C/Si ratio cannot be controlled as easily. It is set by the implantation itself and can be defined as the ratio of the concentrations of interstitial carbon and silicon, generated during the ion bombardment. The concentration of self-interstitials is determined by the implanted fluence and the size and mass of the ions.

At the cost of creating additional damage, the amount of interstitial silicon and carbon as well as the concentration of vacancies can be changed by carbon or silicon co-implantation. Co-implantation has already been successfully employed for III–V compound semiconductors. It could be a suitable tool for forcing the boron atoms to be incorporated mainly on the favoured silicon lattice site.

Experimental examination of boron co-implanted with carbon or silicon has been performed [7,11]. Room-temperature C/B and Si/B implantations were conducted and the concentration of boron acceptors and D-centres was determined by Hall effect and deep level transient spectroscopy (DLTS, see Section 3.2.3.1) measurements. In DLTS spectra, the peak corresponding to the D-centre is clearly observed (FIGURE 3.5). C co-implantation results in a decrease of the D-centre peak height. The peak height is correlated to the centre concentration, indicating a suppressed incorporation of boron as the D-centre.

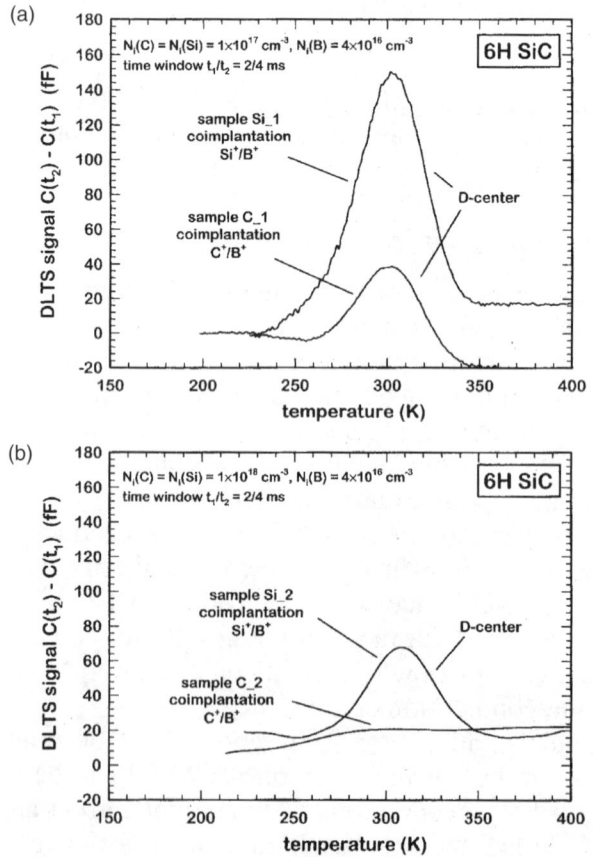

FIGURE 3.5   DLTS spectra of Si/B- and C/B-coimplanted 6H-SiC samples. The shown DLTS peak corresponds to the B-related D-centre. The figures are from [7] and reveal a decrease in peak height for C-coimplantation. The peak height is correlated with concentration of the D-centre.

In the same way, the D-centre concentration is reduced, and an increased concentration of free holes can be observed in Hall effect measurements after C co-implantation (FIGURE 3.6). Although there are still many uncertainties in the literature about the nature of the D-centre, there is strong evidence from double correlated DLTS measurements (see Section 3.2.3.1), that it is acceptor-like. The increase in free hole concentration can be interpreted as being caused by an increased concentration of boron on silicon lattice sites.

Analogous to epitaxial growth, the co-implantation results could be explained by a reduced probability of boron incorporated at carbon sites, when a surplus of carbon competes with boron for vacant lattice sites. Hence, more boron atoms are free to reside on silicon lattice sites. This interpretation would indicate that the microscopic structure of the D-centre includes boron placed on a carbon site. On the other hand, similar results would be obtained if the D-centre involves a carbon vacancy. A surplus of carbon would reduce the concentration of carbon vacancies and also the concentration of the D-centre. Hence, conclusions on the structure of the D-centre cannot be made from the co-implantation experiments. But the investigation clearly demonstrates that carbon co-implantation is a suitable way to force the incorporation of boron on silicon lattice sites and thus increase the concentration of the shallow boron acceptor.

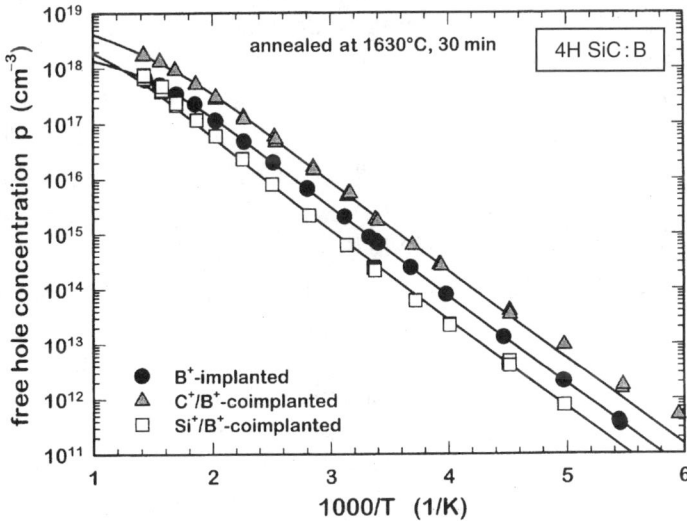

FIGURE 3.6   Free hole concentration p as a function of the reciprocal temperature of three boron implanted 4H-SiC epilayers annealed at 1630 °C for 30 min. The figure is from [7] and reveals an increase (decrease) in the free hole concentration by C-coimplantation (Si-coimplantation) compared to the sample with only B-implantation.

Boron precipitates

Boron-implanted 6H-SiC samples were studied with respect to structural defects by high-resolution transmission electron microscopy (TEM) [12]. The investigated boron profile consisted of an 8-fold box shape profile. The boron concentration was about $2 \times 10^{18}\,cm^{-3}$ and the profile depth about $0.6\,\mu m$. It was shown that after subsequent anneal at $1700\,°C$ in an argon atmosphere, plate-like crystallographic defects are present in the region of the boron implantation. The depth correlation was done by comparing the TEM images with depth profiles observed by secondary ion mass spectrometry (SIMS). The observed defects are associated with a large strain field. The analysis of the strain field and elemental mapping of the defects revealed that the defects contain boron, indicating a precipitate nature, and that they are planar with an extrinsic stacking fault on the 6H-SiC (0001) planes. The crystal planes are bent in order to adjust for the difference in stacking sequence causing the large strain field around the defect. It is expected that the number of plate-like boron precipitates increases with higher boron concentration, an unwanted feature in device structures.

3.2.2.3   Donor dopants in SiC: nitrogen and phosphorus

The group V elements nitrogen, phosphorus, arsenic and antimony form donor levels in SiC. Nitrogen and phosphorus are the donor species that have been investigated in detail and are mostly used for n-type doping in device processing. Nitrogen is used more in epitaxial growth, whereas phosphorus seems to have an advantage in implantation doping. A review on nitrogen doping by implantation can be found in [13] and phosphorus implantation was investigated in [14].

Electrical properties of nitrogen and phosphorus

Nitrogen and phosphorus form the shallowest donor levels known so far in SiC. Nitrogen preferentially occupies carbon lattice sites and phosphorus occupies silicon lattice sites.

Due to the possibilities in stacking silicon–carbon bilayers on top of each other, there exist lattice sites in SiC that have a surrounding layer stacking in hexagonal form and others with cubic form. Sites with hexagonal-structured surrounding are denoted hexagonal sites and those with cubic stacking are denoted cubic sites. Hexagonal and cubic sites differ in the location of the next-nearest neighbour. The cubic 3C-SiC consists only of cubic sites, while the hexagonal 2H-SiC includes only hexagonal sites. In higher-order polytypes, hexagonal and cubic sites exist in the same crystal structure. 4H-SiC has one hexagonal and one cubic site, 6H-SiC one hexagonal and 2 cubic, and the rhombohedral 15R-SiC two hexagonal and three cubic sites. The structure of the SiC polytypes 3C, 2H, 4H, 6H and 15R and the number of hexagonal and cubic lattice sites are shown in FIGURE 3.7. If there exists more than one hexagonal or cubic site in the crystal structure, as in 6H-SiC or 15R-SiC, the surrounding stacking differs first in the 3rd- and 4th-nearest neighbour.

How does the different crystal structure around a lattice site affect the electrical properties of impurities? The donor energy levels arise from the total potential, which is the sum of the host and the impurity potentials. Differences in the position and in the kind of the surrounding atoms lead to differences in the total potential and therefore also in the donor energy levels. The potential influence of a surrounding atom in the crystal is strongly dependent on its distance from the donor site. Atoms far away have no effect on the potential. Because the surroundings of hexagonal and cubic sites differ in the next-nearest neighbour, the difference in the energy levels for donors residing on hexagonal and cubic sites is relatively large. The energy difference for inequivalent cubic or inequivalent hexagonal sites is small, because they differ in the 3rd- and 4th-nearest neighbour.

Contrary to p-type SiC material, where only a small polytype dependence of the acceptor energies is observed, the ionisation energies of nitrogen and phosphorus donors in n-type SiC vary substantially with the SiC polytype. The reason is the difference in the band structure, in particular the location of the conduction band edge and the effective electron mass.

The site and polytype dependences of the ionisation energies for nitrogen and phosphorus have been determined optically and electrically. In low-temperature photoluminescence, for example, a set of transition lines are observed, which are correlated to donors residing on different hexagonal and cubic lattice sites. The number of transition lines corresponds to the number of different hexagonal and cubic lattice sites. In electrical measurements on higher-order polytypes (4H-SiC, 6H-SiC), typically two energy levels per impurity are observed. The energy difference of the two levels is of the order of 30–50 meV. The shallower level is attributed to the impurity residing on hexagonal lattice sites and the deeper one to the impurity on cubic sites. The energy variation of different hexagonal or cubic lattice sites is of the order of a few meV and cannot be resolved in electrical measurements. Instead, a concentration ratio of the two energy levels is determined, which corresponds to the ratio of different hexagonal and cubic lattice sites (e.g. 1 : 2 in 6H-SiC, see FIGURE 3.7).

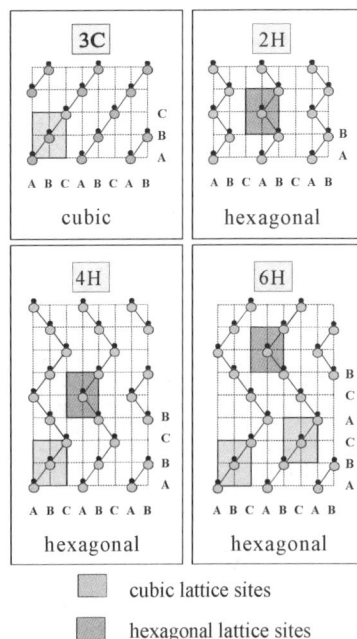

FIGURE 3.7   Bi-layer stacking for the SiC polytypes 3C, 2H, 4H and 6H. The different hexagonal and cubic lattice sites are marked.

TABLE 3.2 Ionisation energies of shallow donor impurities in the semiconductors 4H-SiC, 6H-SiC, silicon, and gallium arsenide [9,13–15]. For SiC two ionisation energies are given, representing electrically observed hexagonal and cubic lattice sites.

Semi-conductor	Shallow acceptor species	Ionisation energy (meV)
4H-SiC	nitrogen	50, 92
	phosphorus	53, 93
6H-SiC	nitrogen	85, 140
	phosphorus	80, 110
Si	phosphorus	45
	arsenic	54
GaAs	silicon	5.8
	germanium	6

TABLE 3.2 summarises the ionisation energies of nitrogen and phosphorus for the hexagonal polytypes 4H-SiC and 6H-SiC and compares them to the ionisation energies of shallow donors in silicon and gallium arsenide. The ionisation energies of nitrogen and phosphorus in SiC do not differ very much and they are not much deeper than the ionisation energies of shallow dopants in silicon.

FIGURE 3.8 shows the calculated temperature dependence of the degree of ionisation for two concentrations of nitrogen and phosphorus doping in 4H-SiC and 6H-SiC. The presence of two donor levels in the material is assumed. The calculation parameters are given in FIGURE 3.8.

Due to the lower values for the phosphorus ionisation energies in 6H-SiC and the concentration ratio of donors residing on hexagonal and cubic lattice sites (6H-SiC: $N_{Dk} = 2 \times N_{Dh}$), doping with phosphorus results in slightly higher values for the degree of ionisation. Hence, the free electron concentration and the conductivity of a phosphorus-doped region are higher. In 4H-SiC the ionisation energies for nitrogen and phosphorus coincide within the measurement errors. No large difference in the degree of ionisation for nitrogen and phosphorus can be observed in the calculations drawn in FIGURE 3.8.

Results in the literature on nitrogen and phosphorus doping by ion implantation are somewhat confusing. For similar implanted concentrations lower resistances are reported for either nitrogen or phosphorus doping and the ionisation energies measured by Hall effect vary in the order of 10 meV. These

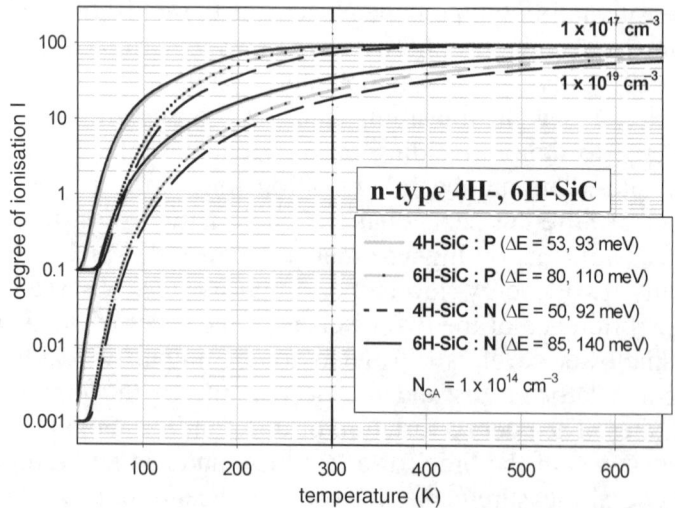

FIGURE 3.8   Calculated degree of ionisation I as a function of the temperature for nitrogen and phosphorus doping of n-type 4H- and 6H-SiC. The calculation parameters are summarised in the figure.

differing results are probably related to the post-implantation anneal process. Incomplete re-crystallisation will lead to remaining electrically active intrinsic defects, which often are of acceptor character and cause compensation. Compensating acceptors are assumed to be totally ionised in n-type material and carry a negative charge. These negative charges create a locally varying potential and affect the ionisation energy of donors by Coulomb interaction. A high concentration of compensating acceptors results in a reduction of the donor ionisation energy.

3.2.2.4 Alternative dopants

In Sections 3.2.2.2 and 3.2.2.3 the main shallow dopants for n- and p-conductive SiC and their behaviour under ion implantation were discussed. It was shown that the donors nitrogen and phosphorus are sufficient for doping SiC with a high degree of ionisation. In p-type SiC, the shallowest acceptor found so far is aluminium with an energy level 200 meV to 240 meV above the valence band. Acceptors form deep levels in SiC and have a low degree of ionisation in the temperature range in which devices are typically operated. To achieve low-resistive p-type SiC, high acceptor concentrations in the range of $10^{20}\,cm^{-3}$ have to be introduced.

The lack of a shallow acceptor in SiC is one of the major driving forces to investigate alternative dopants. In addition, it is important to study impurities that are unintentionally introduced during the doping process with respect to their electrical properties. Unintentionally introduced impurities can harm the performance, reliability and stability of device structures.

Due to the availability of ion sources that can generate ions of almost any stable element in the periodic table, ion implantation is a perfect doping tool for studying the physical properties of impurities. The only disadvantage is that with increasing ion mass the implanted regions become narrow and more damage is created. High-energy accelerators are needed for heavy ions to form doped regions with a depth sufficient for characterisation and devices. Otherwise, there is no limit to increasing the knowledge on doping of SiC and in that way increasing the available toolbox for the design of device structures.

TABLE 3.3 summarises the ionisation energies for a selection of alternative impurities investigated mainly by Hall effect measurements. Unfortunately, a shallower acceptor than aluminium has not yet been found.

The chalcogens sulphur, selenium and tellurium form double donor systems in silicon and germanium. These elements are expected to form several donor levels also in SiC. The observation

Degree of ionisation for two donor levels with ionisation energies ΔE_{Dh} and ΔE_{Dk} and concentration N_{Dh} and N_{Dk} ($N_D = N_{Dh} + N_{Dk}$):

$$I = \frac{N_{Dh}^+ + N_{Dk}^+}{N_D}$$

$$= \frac{1}{1 + \dfrac{g_D n}{N_C}\exp\!\left(\dfrac{\Delta E_{Dh}}{k_B T}\right)}$$

$$+ \frac{1}{1 + \dfrac{g_D n}{N_C}\exp\!\left(\dfrac{\Delta E_{Dk}}{k_B T}\right)}$$

I	degree of ionisation
n	free electron concentration
N_{Dh}^+	concentration of ionised donors on hexagonal sites
N_{Dk}^+	concentration of ionised donors on cubic sites
N_{Dh}	doping concentration of donors on hexagonal sites
N_{Dk}	doping concentration of donors on cubic sites
N_D	doping concentration $N_D = N_{Dh} + N_{Dk}$ of donors
ΔE_{Dh}	ionisation energy of the hexagonal donor
ΔE_{Dk}	ionisation energy of the cubic donor
N_C	effective density of states of the conduction band
g_D	degeneracy factor (set to 2 for donors)
k_B	Boltzmann constant
T	temperature

TABLE 3.3   Ionisation energy and impurity type for alternative impurities in the hexagonal SiC polytypes 4H and 6H. Acceptor energies are given with respect to the valence band ($\Delta E(A) = E(A) - E_V$) and the donor energies with respect to the conduction band ($\Delta E(D) = E_C - E(D)$). The indices h and k indicate the energies for donors residing on hexagonal and cubic lattice sites, respectively.

Element	SiC polytype	Ionisation energy (meV)		Characterisation method	Ref.
ACCEPTORS					
Beryllium	6H	Be	320		[16]
Gallium	4H	Ga	290	Hall effect & admittance	[8]
	6H	Ga	300	spectroscopy	
DONORS					
Sulphur	4H	S_h	275	Hall effect	[17]
		S_k	410		
Arsenic	4H	As_h	67	Hall effect	[18]
		As_k	127		

of two energy levels for sulphur doping of 4H-SiC was related in [17] to substitutional sulphur on hexagonal and cubic lattice sites. However, the energy difference between the two sulphur levels is 135 meV, which is much higher than the energy difference between the nitrogen and phosphorus levels (30–50 meV). Another interpretation could be given in analogy with the observations in silicon. Sulphur has two electrons more in the outer shell than silicon and carbon. Therefore, it can exist in two charge states (S^0, S^+), which give rise to two energy levels in the bandgap.

Arsenic belongs to the same group of elements in the periodic table as nitrogen and phosphorus with one electron more in the outer shell than silicon and carbon. It forms donors and, in analogy to nitrogen and phosphorus, two energy levels in the SiC bandgap are observed by Hall effect measurements [18]. The energy difference of the two arsenic levels is 60 meV and similar to the energy difference for the nitrogen and phosphorus levels, indicating incorporation on hexagonal and cubic lattice sites.

3.2.3   Deep impurities

Besides the shallow impurities (Sections 3.2.1 and 3.2.2), which govern the conductivity type, intrinsic and extrinsic deep defect centres can be present in doped and undoped SiC. Deep defect centres often act as charge trapping centres or recombination centres, limiting the charge carrier lifetime. Hence, the electrical properties of doped and undoped layers are affected by the presence of deep centres and their study is important for the design of electronic devices.

Before we discuss deep impurities in SiC, a short description on how to characterise them follows in the next section.

3.2.3.1   Characterisation of deep impurities

The characterisation of deep impurities in semiconductors is done with electrical, optical and structural methods. The characterisation methods for shallow dopants (see Section 3.2.2.1) can also be applied to deep impurities. SIMS is used to determine the atomic concentration and distribution of deep impurities. The electrical methods, Hall effect and CV, are not sensitive to deep centres. The concentration of deep centres is often lower than the concentration of shallow dopants, so that the signal from deep centres is hidden in the signal obtained from shallow impurities. The electrical method for investigating deep centres is deep level transient spectroscopy (see below).

To identify the structure of deep centres (complex or isolated impurity), methods like electron spin resonance (ESR), electron paramagnetic resonance (EPR), and related methods are employed. With these methods the interaction (spin interaction, hyperfine interaction) between impurities and their neighbouring atoms is studied. From this interaction possible defect structures are determined. For the design of devices, the actual structure of defects is not important. What has to be known are their electrical properties, and if they can harm the functionality or reliability of the device. Hence, energy level, capture cross section for charge carriers, concentration, and impurity type are the parameters to be implemented in simulation tools for the design of devices.

To clearly identify the properties of deep defect centres, it would be beneficial if electrical, optical and structural measurements could be performed on the same set of samples. Unfortunately, the requirements of these methods are so different that a set of samples suitable for all methods is almost impossible to prepare. For example, deep level transient spectroscopy and low-temperature photoluminescence are measured on low-doped samples, whereas ESR and EPR need thick layers with high impurity concentrations and are often measured on bulk material.

Deep level transient spectroscopy

Deep level transient spectroscopy (DLTS) is an electrical measurement method for energetically deep centres in the bandgap of

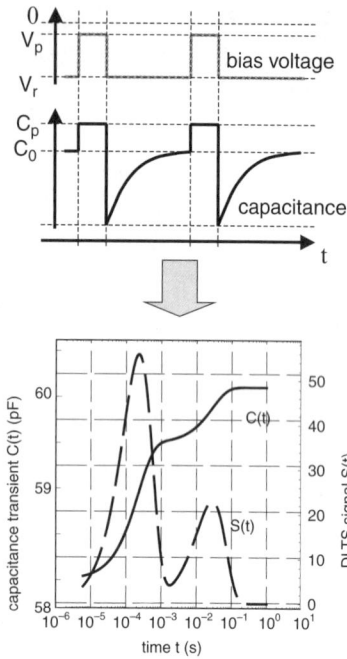

FIGURE 3.9 Principle and spectroscopic analysis of DLTS measurements.

semiconductors. DLTS is based on the Shockley-Read-Hall statistic and exists in a variety of different technological realisations. Common in all DLTS techniques is the detection of a transient caused by the charge carrier emission from deep centres after a filling pulse. This charge carrier emission takes place with a time constant characteristic of a particular deep defect and the observed transients can be analysed spectroscopically. Monitoring the time constant with changing temperature and applying an Arrhenius plot analysis, the ionisation energy and the capture cross section of the deep centre can be determined. The height of the observed transient is a measure of the trap concentration.

The differences between the various DLTS methods are the means of creating the filling pulse (voltage pulse, laser irradiation), and the detected electrical characteristic (capacitance, current). The most common DLTS method is a capacitance method. The investigated area is the space charge region of a Schottky or pn-diode. During a voltage pulse applied to the diode, the deep centres in the space charge region are filled. Applying a reverse bias after the filling pulse causes the deep centres to release their captured charges and a capacitance transient is observed. The basic principle of the capacitance DLTS method is shown in FIGURE 3.9.

The capacitance DLTS method is a very sensitive method. Defect concentrations of the order of a factor 10^5 lower than the shallow dopant concentration can be detected. In Schottky and pn-diodes the region investigated is determined by the shallow doping concentration and consists of the difference between the space charge region under pulse and reverse bias conditions. If the pulse voltage and the reverse bias voltage are not very different, then the investigated sample region is small and the electric field in the diode can be calculated. By scanning the bias conditions with constant difference between pulse and reverse voltage, the deep centre energy is evaluated as a function of the electric field. This method is often referred to as double correlated DLTS (DDLTS). DDLTS is less sensitive than DLTS because the investigated region and the capacitance transient signal are small, but it gives additional information about charge states after the charge carrier emission and impurity type. According to the Poole-Frenkel effect, acceptors in n-type material and donors in p-type material have no field dependence. The field dependence of donors in n-type material and the field dependence of acceptors in p-type material depend on the impurity charge state after carrier emission.

3.2.3.2 Identification of extrinsic and intrinsic impurities

DLTS measurements on implanted samples often result in spectra with many different peaks. It is difficult to reveal whether a peak is related to introduced impurities or to intrinsic defects. One possible method to distinguish between energy levels caused by extrinsic defects and energy levels caused by intrinsic defects is implanting ions from neighbouring elements in the periodic table (e.g. vanadium, titanium, scandium) and comparing the resulting DLTS spectra. For this investigation, a set of identical samples should be prepared with similar implantation dose. The implanted samples are compared with the as-grown sample, to avoid misinterpretation of the results. It is assumed that neighbouring elements cause similar intrinsic defects with nearly the same concentration. In addition, implantation of group VIII elements with a mass similar to the species under investigation provides further results.

Implantation of radioactive isotopes

Another way to distinguish between DLTS peaks from extrinsic and intrinsic impurities is the implantation of radioactive isotopes. The DLTS spectra are monitored as a function of time and changes in the peak height are observed, if radioactive atoms are involved. If there is no change in the spectra, the observed deep centre is either of intrinsic nature or related to non-radioactive atoms. Usually, 100% clean radioactive material for implanter sources is not available and non-radioactive isotopes are implanted together with the radioactive ones.

DLTS measurements last typically 30 min to a few hours, depending on the investigated temperature range. To have sufficient time to carry out the DLTS measurements, the decay time for the radioactive transition should be of the order of several hours to days. Otherwise, the concentration of the defects including the radioactive isotopes is changing during the measurement and the peak analysis leads to false interpretations.

Implantation of radioactive isotopes was successfully employed to determine the energy levels for the transition metals vanadium, titanium, and chromium [19] (see Section 3.2.3.3).

3.2.3.3 Extrinsic deep defect centres

Extrinsic deep defect centres are impurities intentionally or unintentionally introduced during the doping process. In bulk or epitaxial material growth extrinsic deep centres are unintentionally incorporated caused by residual contamination of the growth reactor. Contamination is a smaller problem in ion implantation as accelerators are operated under high vacuum and at room temperature. Electrostatic and magnetic filters effectively select the ions to be implanted. The critical parameter for the filters is the ratio between ion charge and ion mass (q/m-ratio). All ions with the same q/m-ratio pass the filters and are implanted into the crystal. Hence, the ion source material has to be sufficiently clean and should contain only one element with the q/m-ratio to be implanted.

In ion implantation, extrinsic deep centres are introduced intentionally to investigate their properties or to tailor the electrical properties of certain regions in a device structure. TABLE 3.4 summarises the electrical properties of investigated extrinsic deep defect centres.

The transition metals titanium, chromium and vanadium play an important role in SiC, as they are the most frequent residual impurities in SiC crystals. Titanium and chromium have two acceptor levels in 4H-SiC close to the conduction band, which are attributed to titanium and chromium residing on hexagonal and cubic lattice sites. Applying the Langer-Heinrich rule, and noting the 220 meV smaller bandgap of 6H-SiC compared to 4H-SiC, the acceptor levels of titanium and chromium on hexagonal and cubic lattice sites are located in the 6H-SiC conduction band and hence cannot be observed by DLTS or other electrical characterisation methods.

Extrinsic deep centres are intentionally introduced into the material to create regions with high resistivity by compensating

TABLE 3.4 Ionisation energy, capture cross section, and impurity type for a selection of investigated extrinsic impurities in the hexagonal SiC polytypes 4H and 6H. All energies are given with respect to the conduction band. The indices h and k indicate the energies for impurities residing on hexagonal and cubic lattice sites, respectively.

Element	SiC polytype	Ionisation energy $\Delta E = E_C - E$ (eV)		Estimate for the capture cross section (cm^2)	Characterisation method	Ref.
ACCEPTORS						
Vanadium	4H	V_1/V_2	0.9	3×10^{-16}	DLTS	[19,20]
	6H	V_1/V_2	0.65	2×10^{-16}		
		V_3	0.7			
Chromium	4H	$Cr_{1,h}$	0.15	1×10^{-15}	DLTS	[19]
		$Cr_{2,k}$	0.18	1×10^{-15}	+ radio tracer	
		Cr_3	0.74	2×10^{-15}		
	6H	Cr_3	0.54	1×10^{-15}		
Titanium	4H	$Ti_{1,h}$	0.12	5×10^{-15}		[19,20]
		$Ti_{2,k}$	0.16	1×10^{-14}		
DONORS						
Vanadium	4H	V_d	1.42		Hall effect	[21]
	6H	V_d	1.47			

the shallow dopants. This is successfully employed in III–V compound semiconductors, where, for example, iron forms a midgap acceptor level in indium phosphide and compensates the shallow donors. Iron concentrations higher than the donor concentration results in semi-insulating material. In this respect, the transition metal vanadium has recently gained more attention among research groups working with SiC. The vanadium donor was employed in bulk growth and ion implantation to form semi-insulating material. Vanadium is known to have amphoteric character with a midgap donor level (V^{5+}/V^{4+}) and an acceptor level (V^{4+}/V^{3+}).

3.2.3.4 *Intrinsic deep defect centres*

Intrinsic impurities are vacancies, anti-sites, or complexes between them and with interstitial silicon and carbon. In the growth of bulk and epitaxial material the formation of deep intrinsic centres compared to the contamination with extrinsic impurities is small. But still the signature of intrinsic deep centres can be found in results from electrical and optical characterisation methods. In ion implantation, the creation of intrinsic defects is built into the technology. The ion bombardment causes collisions of ions with the crystal atoms, which are kicked out from their sites and end up partly in interstitial sites, leaving a vacancy behind. In compound semiconductors, independent sublattices for each compound are present. Anti-site defects are formed through atoms residing on the wrong sub-lattice site. In SiC, a carbon atom sitting on a silicon site is called a carbon anti-site and vice versa a silicon atom residing on a carbon site is a silicon anti-site.

Intrinsic defects have been investigated in detail in as-grown bulk and epitaxial material and after ion implantation, as well as electron or proton irradiation. Most of the observed defect centres appear in the upper half of the SiC bandgap and are acceptor-like. DLTS after irradiation and anneal below 1000 °C reveals more than 10 different deep centres. Some of these centres are thermally unstable above 1000 °C and anneal out at temperatures commonly used for activation of the implanted dopants. A review on intrinsic defects observed in different SiC polytypes is given in [20]. TABLE 3.5 summarises the electrical properties of the most stable intrinsic defects observed by DLTS. The identification of a DLTS peak as an intrinsic defect was done by either comparing DLTS spectra from samples implanted with neighbouring elements in the periodic table or by helium implantation.

The most prominent intrinsic defect is the Z_1-centre in 4H-SiC and the corresponding Z_1/Z_2-centre in 6H-SiC. The Z_1-centre and

TABLE 3.5   Ionisation energy, capture cross section, and impurity type of a number of intrinsic defect centres observed by DLTS in the hexagonal SiC polytypes 4H and 6H. All energies are given with respect to the conduction band [20].

Defect centre	SiC polytype	Ionisation energy $\Delta E = E_C - E$ (eV)	Estimate for the capture cross section (cm^2)	Observed in
ACCEPTORS				
Z_1 Z_1/Z_2	4H 6H	0.65 0.7	1×10^{-14} 5×10^{-15}	as-grown & implanted
RD_4	4H	1.5	5×10^{-14}	implanted
E_1/E_2	6H	0.42	2×10^{-14}	as-grown implanted
R	6H	1.2	2×10^{-14}	as-grown & implanted

the Z_1/Z_2-centre are reported to be stable at least up to temperatures of 2000°C and 1700°C, respectively. It is suggested that both centres consist of a nearest-neighbour divacancy V_C-V_{Si} and correspond to the D_1-centre observed in luminescence spectra.

In 4H-SiC, as well as in 6H-SiC, midgap intrinsic centres are observed after irradiation, which are denoted RD_4 and R, respectively. These centres are particularly interesting for device applications, because the defect energy is close to the middle of the bandgap and the capture cross section for electrons is of the order of 2×10^{-14} cm^2. These electrical properties make the R- and RD_4-centre efficient recombination centres, which limit the lifetime of free charge carriers. The microscopic structure of these centres is so far not known. Both centres can be observed after anneal at temperatures of 1000°C, but their concentration drops below the detection limit of DLTS after annealing the implanted samples at temperatures above 1400°C.

3.3   DIFFUSION

In silicon, diffusion is an important semiconductor doping process for fabricating devices. For the processing of SiC devices, diffusion plays a minor role due to negligible diffusion coefficients of common doping species and temperatures in excess of 2000°C needed to diffuse dopants to some extent. The consequence of this insignificance is that diffusion data are incomplete and there is little scientific attention on studying diffusion processes in the various SiC polytypes. However, despite the fact that diffusion is not a practical doping process in SiC, diffusion mechanisms have to be investigated as dopants are redistributed during the high-temperature processes of epitaxial growth and post-implantation anneal.

3.3.1 Diffusion for SiC: general process

Most data on solubility limits and diffusion coefficients of doping elements in SiC are available for the polytype 6H in the temperature range above 1800 °C. A thorough investigation of diffusion in other SiC polytypes has not yet been performed. However, it is expected that the migration behaviour of impurities does not vary very much between the SiC polytypes. A summary of solubility limits and diffusion data can be found in [22]. TABLE 3.6

TABLE 3.6   Maximum solubility limit and effective diffusion coefficient at elevated temperatures for various impurities in 6H-SiC and silicon. The data are taken from [9, 22].

Semi-conductor	Doping element	Maximum solubility limit		Effective diffusion coefficient		
		Concentration (cm^{-3})	Temperature (°C)	Fast branch (cm^2/s)	Slow branch (cm^2/s)	Temperature range (°C)
6H-SiC	Nitrogen	6×10^{20}	>2500	–	5×10^{-12}	1800–2450
	Phosphorus	2.8×10^{18}	>2500	–	–	–
	Arsenic	5×10^{16}	>2500	–	–	–
	Oxygen	–	–	–	1.5×10^{-16}–5×10^{-13}	1800–2450
	Aluminium	2×10^{21}	>2500	–	3×10^{-14}–6×10^{-12}	1800–2300
	Boron	2.5×10^{20}	>2500	2×10^{-9}–1×10^{-7}	2.5×10^{-13}–3×10^{-11}	1800–2300
	Beryllium	8×10^{20}	>2500	2×10^{-9}–1×10^{-7}	3×10^{-12}–1×10^{-9}	1800–2300
	Gallium	2.8×10^{19}	>2500	–	2.5×10^{-14}–3×10^{-12}	1800–2300
Si	Phosphorus	1.2×10^{21}	1200	1×10^{-14}–6×10^{-11}	–	1100–1400
	Arsenic	2×10^{21}	1100	5×10^{-16}–6×10^{-12}	–	1100–1400
	Aluminium	2×10^{19}	1200	1×10^{-13}–2×10^{-10}	–	1100–1400
	Boron	6×10^{20}	1400	3×10^{-14}–5×10^{-11}	–	1100–1400

summarises the maximum solubility limits and the diffusion coefficients for a selection of shallow acceptor and donor dopants in SiC and compares them with the equivalent data for shallow dopants in silicon.

In general, the diffusion coefficient for a dopant species and temperature is several orders of magnitude lower in SiC than in silicon. Due to the high band energy and the small interatomic distances in SiC, the diffusion of larger atoms is strongly prohibited and the diffusion coefficients are almost negligible. Only small atoms like hydrogen, lithium, boron and beryllium have a fast diffusion branch and migrate in SiC to a larger extent. The smaller atoms (hydrogen and lithium) migrate mainly via interstitial mechanisms, where the slightly bigger atoms (boron) diffuse mostly through carbon and/or silicon vacancies.

Besides bulk material growth, where temperatures in the range of 2500 °C are applied, SiC device processing is mostly done at temperatures far below 2000 °C. There are only two processes utilising higher temperatures, the epitaxial growth and the post-implantation anneal. Epitaxial growth is typically done in the temperature range of 1300 °C to 1800 °C. In a post-implantation anneal, temperatures in the range of 1500 °C to 2000 °C are applied (see Section 3.2.1). In epitaxial growth, aluminium and nitrogen are commonly used to achieve p-type and n-type conductivity, respectively. A significant diffusion of dopants from the growing epilayer into the substrate is not expected because of the low diffusion coefficients. In the next sections we will therefore concentrate on dopant redistribution during post-implantation anneal and discuss mainly the acceptors aluminium and boron. The donors nitrogen and phosphorus behave similarly to aluminium under high-temperature treatment.

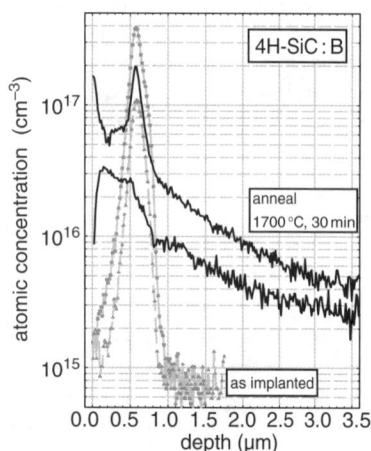

3.3.2 Boron redistribution during implantation anneal

The redistribution of boron during a 1700 °C anneal for 30 min is shown in FIGURE 3.10. Boron was implanted into a 4H-SiC epitaxial layer with peak concentrations of $1 \times 10^{17} \, \text{cm}^{-3}$ and $4 \times 10^{17} \, \text{cm}^{-3}$. The implantation energy was 300 keV at a temperature of 600 °C. The depth profiles in FIGURE 3.10 were measured using SIMS and are typical for as-implanted and annealed boron-doped SiC samples.

In comparison to the as-implanted profile, the boron profile after the implantation anneal at 1700 °C for 30 min is not symmetric. A several micrometre long tail has developed into the bulk of the epitaxial layer starting at a concentration approximately an order of magnitude lower than the peak concentration. An increased boron concentration can also be observed in the

FIGURE 3.10 Depth profiles obtained by SIMS on boron implanted 4H-SiC epilayers before and after a heat treatment at 1700 °C for 30 min in argon atmosphere. The profiles are shown for the peak concentrations of $1 \times 10^{17} \, \text{cm}^{-3}$ and $4 \times 10^{17} \, \text{cm}^{-3}$ with the implant energy of 300 keV and a sample temperature of 600 °C.

region from the profile peak towards the surface of the epilayer. The diffused boron concentration in this region is higher than the diffused concentration into the epilayer with an approximately factor 3 higher starting point at the implanted profile. In the case of the implantation with a peak concentration of $1 \times 10^{17} \, cm^{-3}$, the implanted profile has nearly vanished in the redistribution.

Recently, attempts have been made in the literature to develop a model describing the diffusion mechanism of boron in SiC [23]. The model is based on observed analogies to the boron diffusion in silicon. In silicon, the boron diffusion is governed by the kick-out mechanism. A substitutional boron is kicked out from the lattice site by an interstitial silicon resulting in interstitial boron. In regions damaged by irradiation, an enhancement of the boron diffusion can be observed in silicon. The irradiation causes an excess of silicon interstitials. The silicon interstitial concentration can be several orders of magnitude higher than the silicon interstitial concentration at thermal equilibrium, depending on the irradiation dose.

Kick-out mechanism for diffusion of boron:

$$B_S + Si_I \rightarrow B_I$$

Index S denotes substitutional lattice sites and index I interstitial lattice sites.

The excess of silicon interstitials in regions damaged by irradiation is also the case in SiC. In addition, there are carbon interstitials present. The distribution of silicon and carbon interstitials is disturbed by the irradiation and, due to the heavier mass of silicon, an additional surplus of silicon is accumulated in the region closer to the surface. During the annealing the silicon interstitials become highly mobile and thereby enhance the diffusion of boron towards the surface. The surface acts as a sink for both elements and evaporation can occur according to their vapour pressure.

Similar diffusion behaviour to that for single-energy profiles can be observed in multiple-energy (box-shape) implanted boron profiles. FIGURE 3.11 shows four energy profiles with plateau concentrations of $1 \times 10^{17} \, cm^{-3}$ and $5 \times 10^{18} \, cm^{-3}$, implanted at a temperature of $500 \, °C$ in a high-energy implanter. Towards the surface the diffused boron concentration is enhanced due to the surplus of silicon interstitials. The boron migration into the epilayer is similar to that for single-energy profiles indicating that the source for this diffusion tail is mainly the implantation with the highest energy. For the lower-implanted boron concentration the profile at the surface side is consumed during the annealing at $1700 \, °C$ for $30 \, min$.

The diffusion away from the implanted region is not always desired for device structures. In particular, the outdiffusion at the surface is detrimental to the preparation of ohmic contacts to the surface. Ohmic contacts with low contact resistance require high doping concentrations. Hence, a way has to be found to considerably suppress the boron outdiffusion. This means the

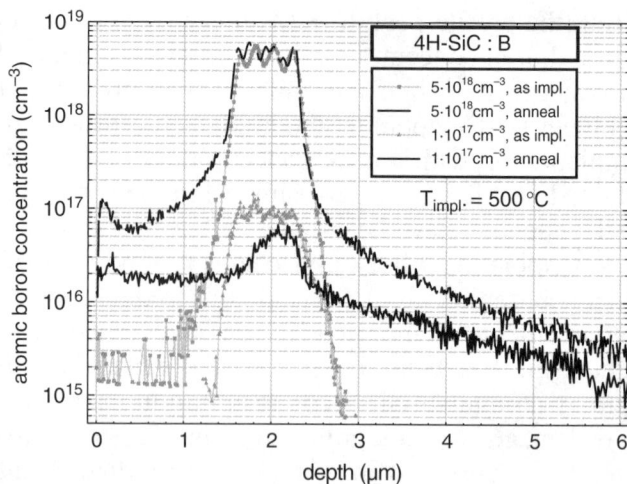

FIGURE 3.11   SIMS depth profiles of boron implanted 4H-SiC n-type epi-layers. Shown are two boron concentrations implanted with 4 energies in the range of 1.4 to 2.2 MeV at a sample temperature of 500 °C. The anneal was done under argon at 1700 °C for 30 min.

concentration of silicon interstitials in the region damaged by the ion implantation has to be strongly reduced before boron starts to move.

It is known from the literature that in 6H-SiC carbon and silicon vacancies recombine with carbon and silicon interstitials at temperatures of 250 °C and 750 °C, respectively [24]. At temperatures below 1000 °C the boron diffusion is almost negligible. A two-step annealing with the first step at 900 °C for at least 2 h and the second step with a temperature of 1700 °C for 30 min was tried and a considerable reduction in the boron diffusion could be observed [25]. Nevertheless, the two-step annealing could not totally prevent the boron from diffusing out at the surface.

Another way of reducing the silicon interstitials is known from silicon. Introducing carbon into silicon binds silicon interstitials by forming stable pairs of interstitial silicon and carbon. All the silicon interstitials bound to carbon no longer participate in promoting the boron diffusion. In SiC, carbon interstitials are always present. An increased carbon interstitial concentration by carbon co-implantation could substantially reduce the amount of interstitial silicon. SIMS investigations of boron/carbon co-implanted SiC show a strongly suppressed migration of boron [25]. The amount of boron migrating into the material is reduced by almost an order of magnitude and the outdiffusion of boron can be reduced by approximately a factor of 2. But also in the case of boron/carbon co-implantation the outdiffusion cannot be totally avoided.

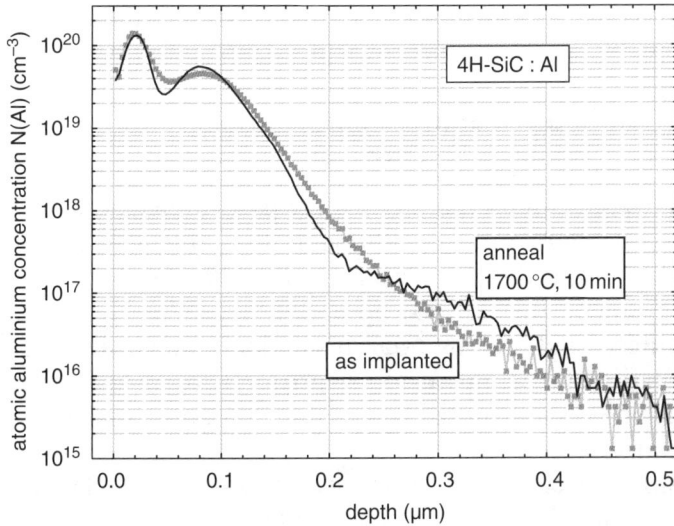

FIGURE 3.12   SIMS depth profiles before and after a heat treatment at
1700 °C for 10 min of aluminium implanted 4H-SiC epilayers. The
implantation was performed with two energies of 30 and 75 keV and a
temperature of 800 °C.

This means that the formation of low-resistive ohmic contacts
to boron-implanted p-type SiC has to involve a dry etching
step to remove the low-doped top layer before the actual metal
deposition.

3.3.3   Aluminium redistribution during post-implantation anneal

The redistribution of aluminium during a high-temperature
treatment is almost negligible. FIGURE 3.12 shows implanted
double-energy profiles before and after anneal at 1700 °C for
10 min. On the bulk side a shoulder can be observed, which is
not present in the profile before the anneal. The maximum of the
second peak is slightly increased and the dip between the first
and the second peak is reduced. No outdiffusion of aluminium
on the surface is observed.

3.4   RECENT DEVELOPMENTS AND FUTURE TRENDS

The recent developments and future trends for ion implantation
concentrate mainly on making the post-implantation anneal
more efficient with better quality, better knowledge of residual
damage and the corresponding electrically active defects, identification of intrinsic and extrinsic defects, and higher dopant

activation. In the following paragraphs ideas for improving the ion-implantation process are discussed.

3.4.1 Rapid thermal processing for the post-implantation anneal

Furnace annealing as the process for post-implantation anneal is a time-consuming process. Although the actual anneal time is as short as 10 min, the heating up is an additional 10 min and the cooling down time can last up to 1 h, depending on the size of the furnace and the thermal insulation. Therefore, the post-implantation anneal process can last for 1 to 1.5 h. Of course, it is possible to build furnaces in which a larger number of wafers can be annealed at a time,

In silicon technology, rapid thermal processes (RTP) utilising high-intensity lamps or lasers are widely used, because silicon is absorbing in the visible range. It would be beneficial for the production of SiC devices if the established RTP processes could be used for thermal treatment. Currently there are investigations going on to establish RTP processes utilising high-intensity flash lamps or pulsed excimer lasers for the post-implantation annealing of SiC [26,27]. Due to its high bandgap of over 3 eV, SiC is transparent and therefore absorbs very little in the visible range. Hence, to anneal SiC in RTP systems initially made for silicon processing the samples have to be placed on absorbing materials or into a graphite container. In addition, when anneal temperatures are higher than 1300 °C or when the annealing is done under high vacuum conditions, special care in the form of cap layers has to be taken to avoid degradation of the surface by silicon evaporation.

The initial experiments with pulsed excimer lasers showed that a high number of shots (in the range of 10000–100000) have to be performed to result in a sufficient annealing of implanted SiC layers [27]. This gives a duration of the actual anneal of around 10 min, which is similar to the furnace anneal. But the heating and cooling can be done within seconds in such laser systems. The activation efficiency of laser annealed aluminium implanted SiC was estimated to be about 0.1%.

Flash-lamp annealing has been applied in [26]. The annealing behaviour of aluminium and nitrogen implanted 6H-SiC was studied. An enhancement of the hole concentration was obtained in high-concentration aluminium-doped layers after flash-lamp annealing at about 2000 °C compared to furnace anneal. Flash-lamp annealing results in no additional damage and the dopants do not diffuse out from the implanted profile. Flash-lamp annealing systems could be built with large active areas, suitable as production equipment.

3.4.2   Channelled implants

Typically, ion implantation is done into as-grown wafers that are not specially oriented (off-axis orientation). This is due to the fact that high-quality epitaxial layers, mainly used for investigations, are grown on off-axis wafers. The off-angle is dependent on the polytype and 6H-SiC is commercially available with an off-angle of 3.5° and 4H-SiC of 8°. Hence, the incoming ion beam sees a random distribution of the semiconductor atoms and collisions with the atoms are not avoidable.

The resulting profiles were described in Sections 3.2.2 and 3.2.3. If the crystal is oriented in one of the main crystallographic directions, like the $\langle 0001 \rangle$ direction, the crystal is no longer densely packed with atoms and channels are open for the incoming ion beam, where the ions can travel without collisions with the crystal atoms. It is obvious that the number of collisions with crystal atoms should be reduced by these channelled implants and the maximum depth the ions can reach is increased.

Channelled implants were demonstrated in [28]. Aluminium, boron and gallium ions were implanted into off-axis and $\langle 0001 \rangle$ oriented 6H-SiC samples with ion energy 1.5 MeV, 1.44 MeV and 3 MeV, respectively. The resulting profiles are shown in FIGURE 3.13. The implanted single-energy aluminium profile is widened by the channelling in the oriented case at least by a factor of 3 as compared to the equivalent random implants. With an aluminium fluence of around $1 \times 10^{13}\,cm^{-2}$, an almost squared profile with nearly uniform concentration of about $1 \times 10^{17}\,cm^{-3}$ and a width of 3 μm can be achieved with a single-energy implantation. Hence, homogeneous doped box profiles can be created using fewer implantation energies.

Due to fewer collisions between the implanted ions and the crystal atoms in the channelled implant case, less damage is created. It was shown in [28] that for ion fluence below $5 \times 10^{14}\,cm^{-2}$ the integral damage can be reduced by a factor of 2.5. It is expected that lower temperatures could be used for the post-implantation anneal process.

There are major technological difficulties to be overcome with the channelled implant method. First, the wafers have to be aligned. This alignment could be directly done in implanters, if they are equipped with a chamber for Rutherford backscattering (RBS) measurements. In RBS measurements, the material is irradiated with helium ions and the distribution of the backscattered helium ions is detected. The channelled orientation is found by minimising the backscattering yield. The alignment of the wafers should be carried out as fast as possible in order to minimise the damage created by the helium ion irradiation.

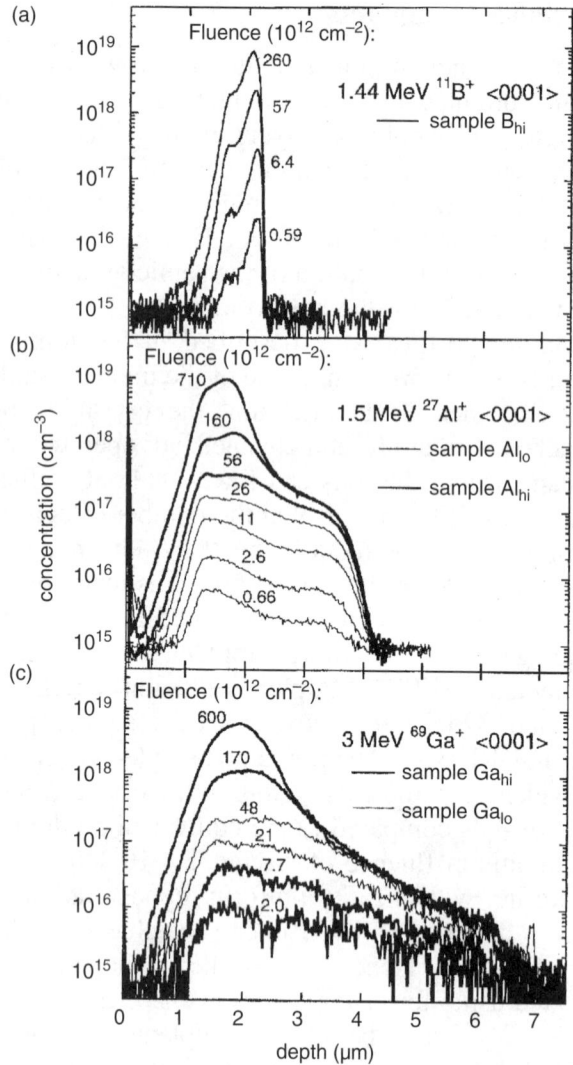

FIGURE 3.13   SIMS depth profiles of boron, aluminium and gallium implanted 6H-SiC oriented in the $\langle 0001 \rangle$ 6H-SiC direction. The figure is taken from [27].

A misorientation as small as about $2°$ prevents the channelling from working properly. To implant large areas the ion beam is usually scanned by electromagnetic deflection over the wafer. To implant large-area wafers, with channelled implantation the ion beam should not be scanned. Instead the wafer should be moved and the ion beam stays in focus.

3.5   CONCLUSIONS

Due to the low diffusion coefficients of the main shallow dopants in SiC, a diffusion process like the one in conventional semiconductor processing is not an applicable process for doping of SiC. Nevertheless, the thorough investigation of diffusion processes in SiC is important, because high-temperature anneal processes have to be used and can result in a redistribution of the dopants. This redistribution has to be taken into account in the design of devices made in SiC.

The most practical doping process for SiC is ion implantation, because of its advantages in selective doping of the material by masking techniques. The ion-implantation process consists of the ion bombardment and a high-temperature annealing at temperatures of around 1700 °C for about 10 min. The high-temperature anneal is needed to activate the implanted species and reduce the damage caused by the ion bombardment.

Ion-implantation doping with the donor species nitrogen and phosphorus, governing the conductivity of n-type SiC, is possible and the conductivity can be varied over a wide range. The available post-implantation anneal is sufficient for activating the introduced donors and residual damage caused by the ion bombardment is low.

To form highly conductive p-type SiC the introduced acceptor concentration has to be very high due to the deep energy levels of boron and aluminium, the most commonly used acceptor dopants in SiC. Hence, the created damage in the crystal is substantially larger than in the n-type case and difficult to remove during the post-implantation anneal. The low conductivity and poor acceptor activation is still a severe problem in the fabrication of SiC devices.

The high-temperature post-implantation anneal process is not yet optimised and improvements can be expected in the near future. Flash-lamp annealing is supposed to be a promising technique to reduce the duration of the anneal process.

REFERENCES

[1] R.I. Scace, G.A. Slack [in *Silicon Carbide – A High Temperature Semiconductor* Eds J.R. O'Conner, J. Smiltens (Pergamon Press, 1960) p.306]

[2] L. Muehlhoff, W.J. Choyke, M.J. Bozac, J.T. Yates [*J. Appl. Phys. (USA)* vol.60 (1986) p.2842–53]

[3] K.A. Jones et al [*J. Appl. Phys. (USA)* vol.83 (1998) p.8010–5]

[4] J.A. Edmond, R.F. Davis, S.P. Withrow [*Ceram. Trans.* vol.2 (1998) p.479]

[5] G. Wellenhöfer, U. Rössler [*Phys. Status Solidi B (Germany)* vol.202 (1997) p.107–22]

[6] G. Rutsch, R.P. Devaty, D.W. Langer, L.B. Rowland, W.J. Choyke [*J. Appl. Phys. (USA)* vol.84 (1998) p.2062–4]

[7] T. Troffer et al [*Phys. Status Solidi A (Germany)* vol.162 (1997) p.277–98]

[8] T. Troffer et al [*Mater. Sci. Forum (Switzerland)* vol.264–268 (1998) p.557–60]

[9] S.M. Sze [in *Physics of Semiconductor Devices*, 2nd edn (John Wiley & Sons, 1981) ch.1, p.20–2]

[10] M. Bockstedte, A. Mattausch, O. Pankratov [*Mater. Sci. Forum (Switzerland)* vol.353–356 (2001) p.447–50]

[11] H. Itoh, T. Troffer, G. Pensl [*Mater. Sci. Forum (Switzerland)* vol.264–268 (1998) p.685–8]

[12] P.O.Å. Persson et al [*Mater. Sci. Forum (Switzerland)* vol.264–268 (1998) p.413–6]

[13] T. Kimoto, N. Inoue, H. Matsunami [*Phys. Status Solidi A (Germany)* vol.162 (1997) p.263–76]

[14] T. Troffer, C. Peppermüller, G. Pensl, K. Rottner, A. Schöner [*J. Appl. Phys. (USA)* vol.80 (1996) p.3739–43]

[15] M.A. Capano, J.A. Cooper, M.R. Melloch, A. Saxler, W.C. Mitchel [*Mater. Sci. Forum (Switzerland)* vol.338–342 (2000) p.703–6]

[16] A.A. Kalnin, V.V. Pasynkov, Y.M. Tairov, D.A. Yaskov [*Sov. Phys.-Solid State (USSR)* (English translation) vol.8 (1967) p.2381]

[17] Y. Tanaka et al [*Mater. Res. Soc. Symp. Proc. (USA)* vol.622 (2000) p.T8.6.1–6]

[18] J. Senzaki et al [*Mater. Res. Soc. Symp. Proc. (USA)* vol.622 (2000) p.T6.7.1–5]

[19] N. Achtziger, J. Grillenberger, W. Witthuhn [*Mater. Sci. Forum (Switzerland)* vol.264–268 (1998) p.541–4]

[20] T. Dalibor et al [*Phys. Status Solidi A (Germany)* vol.162 (1997) p.199–225]

[21] W.C. Mitchel et al [*Mater. Sci. Forum (Switzerland)* vol.264–268 (1998) p.545–8]

[22] G.L. Harris [in *Properties of Silicon Carbide*, Emis Datareviews Series No. 13, Ed. G.L. Harris (INSPEC, IEE, London 1995) ch.7.1, p.153–6]

[23] H. Bracht, N.A. Stolwijk, M. Laube, G. Pensl [*Appl. Phys. Lett. (USA)* vol.77 (2000) p.3188–90]

[24] H. Itoh, M. Yoshikawa, I. Nashiyama, S. Misawa, H. Okumura, S. Yoshida [*J. Electron. Mater. (USA)* vol.21 (1992) p.707]

[25] M. Laube, G. Pensl, H. Itoh [*Appl. Phys. Lett. (USA)* vol.74 (1999) p.2292–4]

[26] D. Panknin, H. Wirth, W. Anwand, G. Brauer, W. Skorupa [*Mater. Sci. Forum (Switzerland)* vol.338–342 (2000) p.877–80]

[27] Y. Hishida, M. Watanabe, K. Nakashima, O. Eryu [*Mater. Sci. Forum (Switzerland)* vol.338–342 (2000) p.873–6]

[28] M.S. Janson et al [*Mater. Sci. Forum (Switzerland)* vol.338–342 (2000) p.889–92]

Chapter 4

Wet and dry etching of SiC

S.J. Pearton

4.1 CHAPTER SCOPE

In this chapter we will discuss wet and dry patterning techniques for SiC and the relative merits of these methods. We describe the basic principles involved in etching SiC and problems that can arise because of the binary nature of the lattice and its relatively high bond strength. Recent developments in the use of high-density plasma sources to achieve fast etching rates (in some cases over $1\,\mu m\,min^{-1}$ for bulk 4H-SiC) are discussed: these sources are likely to play a dominant role for processing of SiC devices since they are capable of producing etch depths from 0.1 to $100\,\mu m$ with minimal disruption of the SiC surface. These processes are also applicable to microelectromechanical systems engineering based on SiC substrates.

4.2 WET ETCHING

Due to its hardness ($H = 9^+$), SiC is one of the most widely used lapping and polishing abrasives for metals, metallic components and semiconductor wafers. However, this very property makes it difficult to etch in typical acid or base solutions. In its single crystal form, SiC is not attacked by single acids at room temperature. Indeed the only techniques for etching SiC employ molten salt fluxes, hot gases, electrochemical processes or plasma etching [1,2]. TABLE 4.1 shows a list of the molten salt solutions and the temperatures needed for successful etching of SiC. The disadvantages of these high-temperature, corrosive mixtures include the need for expensive Pt beakers and sample holders (which can withstand the molten salt solutions) and the inability to etch masked samples because few masks stand up to these mixtures. While one can conceivably use Pt masks, the wet etching is isotropic and therefore undercuts the mask.

TABLE 4.1   Molten flux and other etches for SiC.

Solution	Material	Temperature (°C)	Ref.
NaF/K$_2$CO$_3$	SiC(0001)	650	[3]
H$_3$PO$_4$	a-SiC(H)	180	[4]
NaOH	SiC(111)	900	[5]
Na$_2$O$_2$	SiC(0001)	>400	[6]
NaOH/Na$_2$O$_2$	SiC(0001)	700	[7]
Borax/Na$_2$CO$_3$	epi-SiC	855	[8]
NaOH/KOH	bulk 6H	480	[9]
Na$_2$O$_2$/NaNO$_2$	bulk 6H	>400	[10]
KOH/KNO$_3$	bulk 6H	350	[11]

Photoelectrochemical etching can be successfully employed for SiC [12]. The dissolution rate of semiconductors may be altered in acid or base solutions by illumination with above-bandgap light. The mechanism for photo-enhanced etching involves the creation of e-h pairs, the subsequent oxidative dissociation of the semiconductor into its component elements (a reaction that consumes the photo-generated holes) and the reduction of the oxidizing agent in the solution by reaction with the photo-generated electrons. Generally, n-type material is readily etched under these conditions, while p-type material is not, due to the requirements for confining photo-generated holes at the semiconductor/electrolyte interface (i.e. the p-surface is depleted of holes because of the band bending). This allows for selective removal of n-SiC from an underlying p-SiC layer [12]. Under conditions of no illumination, it is often possible to get the reverse selectivity if the sample is correctly biased, since n-SiC requires photogeneration of carriers for etching to proceed. Etching over large areas can be achieved using Hg lamps and some degree of anisotropy is obtained because of the shadowing effect of the metal masks (typically Ti) allowing carriers to be generated only in unmasked regions. Some of the disadvantages of the technique include fairly rough surface morphologies (due to enhanced dissolution rates for areas around crystal defects), inability to pattern very small dimension features and poor uniformity of etch rate.

4.3   DRY ETCHING

In order to etch silicon carbide in a plasma reactor, the chemistry used must be reactive with SiC and the species produced by the

TABLE 4.2   Published plasma etch rates of SiC.

Reactor	SiC	Gas	Condition at highest ER	ER (Å min^{-1})	Ref.
RIE	6H	CHF_3/O_2	20 sccm, 200 W, 0% O_2	32	[13]
		SF_6/O_2		410	
		CF_4/O_2		278	
		NF_3/O_2		493	
	3C	SF_6	150 W, 80 mtorr	700	[14]
	6H,4H	NF_3	225 mtorr, 95–110 sccm, 275 W	1500	[15]
ECR	6H	CF_4/O_2	5 W, −100 V, 17.5% O_2, 50 sccm	800	[16]
	3C,6H	CF_4/O_2	650 W, −100 V, 17% O_2, 50 sccm	700	[17]
	6H	$20SF_6/10Ar$	750 W, 250 RF, 2 mtorr, 30 sccm	4500	[18]
	3C,5H	SF_6/O_2	1200 W, 1 mtorr, 4 sccm	2500	[19]
	6H	$10Cl_2/5Ar$	1000 W, 150 RF, 1.5 mtorr, 15 sccm	2500	[20]
		$10Cl_2/5H_2$	1000 W, 150 RF, 1.5 mtorr, 15 sccm	1000	
		$4IBr/4Ar$	1000 W, 250 RF, 1.5 mtorr, 15 sccm	1100	
		NF_3	800 W, 100 RF, 1 mtorr, 10 sccm	1600	
		SF_6	800 W, 100 RF, 1 mtorr, 10 sccm	450	
ICP	6H	Cl_2Ar or $/He$	100 Cl_2, 750 W, 250 RF, 5 mtorr	100	[18]
		Cl_2/Xe	13% Cl_2, 750 W, 250 RF, 5 mtorr	260	
		IBr/Ar	10% Ar, 750 W, 250 RF, 5 mtorr	800	
		ICl/Ar	66% Ar, 750 W, 250 RF, 5 mtorr	250	
	6H	NF_3/O_2 or Ar	100% NF_3, 750 W, 250 RF, 5 mtorr	4000	[21]
	4H	NH_3	100% NF_3, 500 W, 50 RF, 2 mtorr	8000	[22]
	6H	SF_6	100% SF_6, 900 W, −450 V, 5 mtorr	9700	[23]
Helicon	4H	$30SF_6/7/5\ O_2$	25% O_2, 200 W, −500 V, 6 mtorr	13.500	[24]

chemical reactions must be volatile compounds under the operating temperature and pressure conditions to avoid residues on the surface.

Many plasma chemistries have been examined (see TABLE 4.2). The most effective gases in terms of etch rate are based on fluorine chemistry. The reaction mechanism of SiC in F_2-based chemistry is shown as below.

$$Si + xF \rightarrow SiF_x \quad x \leq 1\text{–}4$$
$$C + xF \rightarrow CF_x \quad x \leq 1\text{–}2$$

From optical emission spectra it is clear that ion bombardment plays a role in the etch mechanism. When etching silicon atoms with atomic fluorine, a carbon layer is present on the exposed surface and is removed by the ion bombardment.

Various gas additions can have effects on the etch behaviour. Oxygen has often been added to fluorine-based chemistries under RIE conditions to enhance the active fluorine concentration and increase the SiC etch rate. In high ion density conditions, this produces only a small change in the atomic fluorine concentration. In contrast, the addition of H_2 to the gas mixture reduces the etch rate. The introduction of hydrogen into the plasma prevents residue formation through a combination of mechanisms, including the formation of volatile alane (AlH_3) to remove Al sputtered from the reactor and the removal of the C-rich surface.

The differences in the etch rates are due more to differences in the dangling bond densities and the corresponding reactivities of the crystal faces than to the different crystal structures. For example, each atom on a cubic (001) face has two dangling bonds, whereas only one dangling bond exists on a (111) face or similarly on the (0001) face of hexagonal SiC.

Previous results on reactive ion etching of SiC have generally employed F_2-based plasma chemistries, such as CHF_3, SF_6, CH_4 and NH_3, with resultant etch rates $\leq 1500\,\text{Å}\,\text{min}^{-1}$. Relatively rough surfaces are often observed under these conditions due to sputtering of the electrode material onto the SiC sample, leading to micromasking.

With the advent of high-density plasma sources, including electron cyclotron resonance (ECR), inductively coupled plasma (ICP) and helicon, much higher SiC etch rates have been reported. The key advantage of these sources is decoupling of ion energy and ion flux, so that relatively low ion energies can be employed. Schematics of RIE and ICP reactors are shown in FIGURES 4.1 and 4.2. This reduces the electrode-sputtering problem and, in addition, the plasma chemistries for high-density sources generally involve gases that do not contain CH_x because of the extensive polymer deposition that can occur within the source at high applied powers. The absence of these two sources of redeposition onto the SiC generally leads to good surface morphologies.

Inductively coupled plasma etching offers an attractive high-density plasma technique where plasmas are formed in a dielectric vessel encircled by an inductive coil into which RF power is applied. A strong magnetic field is induced in the centre of the chamber, which generates a high-density plasma ($\sim 5 \times 10^{11}\,\text{cm}^{-3}$) due to the circular region of the electric field that exists concentric with the coil. The electrons in this circular path will have only a small chance to be lost to the chamber walls, resulting in low DC self bias. At low pressures ($\leq 20\,\text{mtorr}$), the plasma diffuses from the generation region and drifts to the substrate at relatively low ion energy ($< 25\,\text{eV}$). Thus, ICP etching is expected to produce low damage while

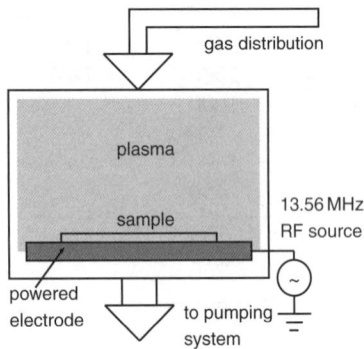

FIGURE 4.1   Schematic of RIE reactor.

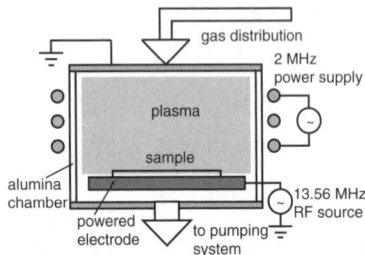

FIGURE 4.2   Schematic of ICP reactor.

achieving high etch rates. Anisotropic profiles are obtained by superimposing an RF bias on the sample to independently control ion energy and by using flow-pressure conditions to minimize ion scattering and lateral etching. ICP sources may be easier to scale up than ECR sources and are more economical in terms of cost and power requirements.

In choosing the optimum plasma chemistries for investigation, it is instructive to look at SiC etch rates reported previously in the literature (TABLE 4.2). There are two key points evident in these data. First, the high-density reactors do indeed produce faster rates, and secondly F_2-based chemistries lead to higher rates than Cl_2, F_2 or Br_2. This is readily understood by examining the relative volatility of the SiC etch products in F_2- or Cl_2-based plasmas. TABLE 4.3 shows the boiling points for potential etch products in these plasmas (with the addition of O_2 in both cases, although it is reported that O_2 itself plays no direct role in SiC etching but rather can influence the etch rate through changing the atomic fluorine neutral density in the discharge). While it is understood that the high ion flux in ICP discharges can desorb the etch products before they are fully coordinated, the boiling points of the complete molecules do give some indication of relative volatility, and hence the trend expected for etch rates in the different chemistries. From TABLE 4.3, it is clear that the fluorinated products are more volatile than their chlorinated counterparts.

FIGURE 4.3 shows a schematic of a through-wafer via for a SiC power transistor. The substrate would generally be thinned to 50–100 μm and it is desirable that the via have a slightly sloped sidewall profile to allow complete step coverage by subsequently deposited metal. Finally, the etching should have a high selectivity over both the mask material and the front-side metal employed as the etch-stop. The thickness of the SiC substrate enables us to estimate that an etch rate of at least $4000 \, \text{Å} \, \text{min}^{-1}$ is needed to keep the process time below ~2 h, which is a rough guess for a practical process.

FIGURE 4.4 shows the etch rates (top) and etch yields (bottom) for SiC in ICP discharges of NF_3, SF_6, BF_3 or PF_5 at fixed RF chuck power (250 W) and pressure (2 mtorr), as a function of ICP source power. The yield tends to decrease as the source power is increased, even as etch rate increases with NF_3, SF_6 and PF_5. This suggests that ion flux is not the limiting factor under these conditions, but rather the supply of fluorine neutrals to the SiC surface limits the etch rate. The etch rates are significantly higher with NF_3 and SF_6, which is consistent with the lower bond strength of these molecules compared to PF_5 and BF_3. When comparing the relative advantages of NF_3 and SF_6,

TABLE 4.3 Boiling points of potential etch products in plasma etching of SiC.

Etching product	Boiling points (°C)
$SiCl_4$	57.6
SiF_4	−86
CCl_4	76.8
CF_4	−128
CO_2	−78.5 subl
CO	−191.5

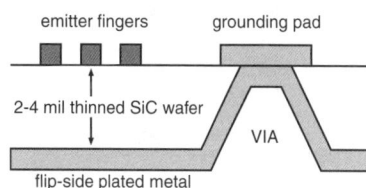

FIGURE 4.3   Schematic of through-wafer via in SiC.

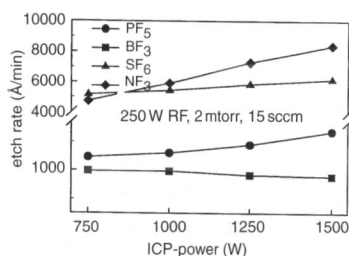

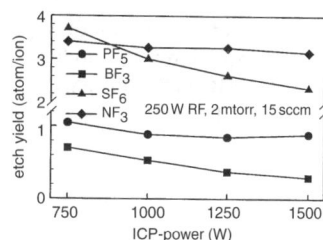

FIGURE 4.4   Etch rates (top) and etch yield (bottom) for SiC in different ICP discharges, as a function of source power.

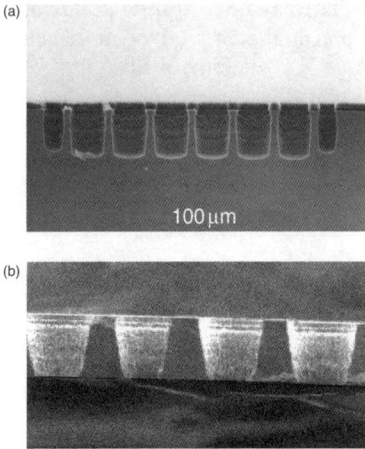

FIGURE 4.5   SEM micrographs
of deep features etched into SiC
substrates using SF_6 discharges.

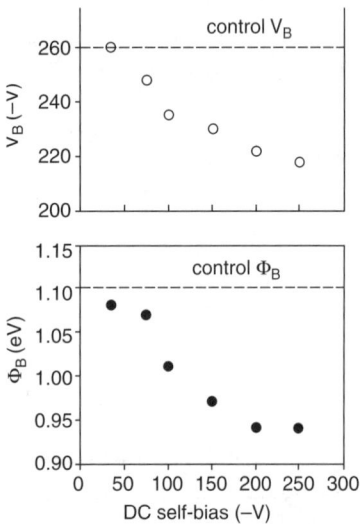

FIGURE 4.6   Variation of V_B and
ϕ_B for n-SiC Schottky diodes
exposed to ICP SF_6 discharges
at different DC self-biases.

the much lower cost of the latter outweighs the faster rates
obtained with the former, particularly for long etch times.

FIGURE 4.5 shows scanning electron micrographs of features
etched ~60 μm (top) or 100 μm (bottom) into SiC substrates.
The top micrograph shows the effect of feature diameter in etch
depth – the smaller diameter features (~30 μm) are shallower by
~15% than the larger openings, which gives the magnitude of
the aspect-ratio-dependent etch rate. The bottom micrograph
shows features etched all the way through 100 μm thick SiC
substrates mounted on sapphire substrates.

In situations in which only a mesa etch is required, it is
desirable that the pattern-transfer process should not degrade the
electrical properties of the SiC. FIGURE 4.6 shows the change
in diode reverse breakdown voltage V_B and Schottky barrier
height (ϕ_B) as a function of the DC self-bias during SF_6 etching
at a source power of 750 W. The decrease in both parameters is
minimal provided the self-bias is kept below ~50 V, corresponding to ion energies of ~75 eV. Under these conditions, the SiC
etch rate is still ~2500 Å min^{-1}. If higher rates are desirable,
then the majority of the etching can be performed at higher DC
self-biases and this latter parameter can be decreased toward the
end of the process.

It is also desirable that there is high selectivity for etching SiC
over the mask material (and also the front-side metallization in
the case of via holes). FIGURE 4.7 shows the dependence of
SiC etch rate (top) and selectivity for SiC over Al (bottom) as
a function of O_2 percentage (by flow) in 500 W source power,
150 W RF chuck power, SF_6/O_2 discharges. Note that the SiC
etch rate initially increases as O_2 is added to the SF_6. This is
probably due to the increase of atomic fluorine neutrals present
at low O_2 percentages, a feature well established for CF_4/O_2
and SF_6/O_2 plasma chemistries. The etch rate falls off at higher
O_2 percentages because atomic oxygen does not appear to play
an active or direct role in etching of SiC. However, the etch
selectivity over Al increases rapidly with O_2 addition, since
the Al oxidizes and does not etch beyond ~40% O_2 addition to
the SF_6.

The fact that ion energy is a key factor in determining the SiC
etch rate is evident from the data of FIGURE 4.8, which shows
the influence of DC self-bias on the etch rate. At fixed source
power, the incident ion energy is controlled by the sum of this
DC self-bias and the plasma potential (-20 to 25 V in this particular tool). The etch rates are always slightly higher with
SF_6/O_2 (25% O_2 by flow in this case) compared to pure SF_6 and
the rates begin to saturate beyond ~350 V where Si-C bond
breaking is no longer the limiting step.

We should also mention that passing hot gases such as Cl_2, F_2, H_2 and HCl over SiC at high temperatures ($>1200\,°C$) will etch the surface and this process is often employed to clean SiC substrates prior to epitaxial growth.

4.4 RECENT DEVELOPMENTS AND FUTURE TRENDS

It has also been recently shown that the use of UV illumination during plasma etching in Cl_2-based gas chemistries can enhance the etch rates of SiC, probably through photo-excitation of the chlorinated etch products. This process does not produce any increase in etch rates with F_2-based gas chemistries, because the etch products are already quite volatile.

The achievement of high etch rates for SiC in the various high-density plasma sources has now placed the emphasis on developing mask materials that can withstand long plasma exposures, such as those needed during via hole formation. The Al masks described earlier work well most of the time, provided the residual stress in the metal is minimized. However, to pattern smaller features, one would ideally like to avoid thick metal masks and use more convenient materials such as photoresists or dielectrics. Unfortunately, these materials etch more rapidly than SiC in F_2-based plasmas, limiting their application to the etching of shallow features.

Since it is clear that more dissociated plasmas with separate control of ion energy produce the fastest etch rates for SiC, it is likely that even higher source powers will be employed in future. Most of the etching to date has been carried out at source powers $\leq 1500\,W$, but reactors are available with powers of 3–5 kW. The higher ion fluxes in these systems will place even greater demands on the durability of mask materials.

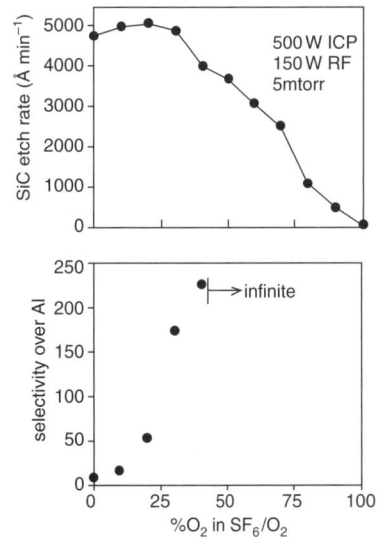

FIGURE 4.7 SiC etch rate (top) and selectivity for SiC-over-Al (bottom) as a function of O_2 percentage in SF_6/O_2 discharges (500 W source power, 150 W RF chuck power, 5 mtorr).

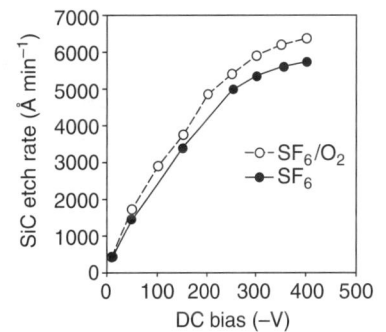

FIGURE 4.8 SiC etch rate versus DC self-bias in SF_6 of SF_6/O_2 (25% O_2 by flow rate), ICP discharges.

4.5 CONCLUSIONS

An ICP process based on SF_6 or NF_3 provides practical etch rates for deep patterning of SiC. The use of the former gas is probably favoured due to its much lower cost and the simpler, less-expensive regulators required. Other F_2-based plasma chemistries involving PF_5 or BF_3 do not produce adequate SiC etch rates. Through-wafer vias have been demonstrated using the ICP SF_6 process, as well as low-damage conditions for etching of mesas. More conventional RIE techniques can also be employed in most situations in device processing, but suffer from lower etch rates and poorer surface morphologies.

REFERENCES

[1] P. Walker, W.H. Tarn (Eds) [*CRC Handbook of Metal Etchants* (CRC Press, Boca Raton, FL, 1991) p.1092–101]

[2] G.L. Harris (Ed.) [in *Properties of SiC*, EMIS Datareviews Series No. 13 (INSPEC, London, UK, 1995) p.134–5]

[3] D. Buckley [*J. Vac. Sci. Technol. A (USA)* vol.3 (1985) p.762]

[4] T.L. Chu, R.B. Campbell [*J. Electrochem. Soc. (USA)* vol.112 (1965) p.955]

[5] W.K. Liebmann [*J. Electrochem. Soc. (USA)* vol.12 (1964) p.885]

[6] E.D. Wolley [*J. Appl. Phys. (USA)* vol.37 (1966) p.1588]

[7] L.B. Griffith [*J. Phys. Chem. Solids (USA)* vol.27 (1966) p.257]

[8] R.W. Brander [*J. Electrochem. Soc. (USA)* vol.12 (1964) p.881]

[9] P. Nordquist, H. Lessoff, R.J. Gorman, M.L. Gripe [*Springer. Proc. Phys. (Germany)* vol.43 (1989) p.119]

[10] P. Pirouz, C.M. Chorey, J.A. Powell [*Appl. Phys. Lett. (USA)* vol.50 (1987) p.221]

[11] M.W. Jepps, T.F. Page [*J. Microsc. (USA)* vol.124 (1981) p.227]

[12] J.S. Shor [in ref. [2], p.141–9]

[13] P.H. Yih, A.J. Steckl [*J. Electrochem. Soc. (USA)* vol.142 (1996) p.312]

[14] J. Wu, J.D. Parsons, D.R. Evans [*J. Electrochem. Soc. (USA)* vol.142 (1995) p.669]

[15] J. Casady, E.D. Luckowski, M. Bozack, B. Sheridan, R.W. Johnson, J.R. Williams [*J. Electrochem. Soc. (USA)* vol.143 (1996) p.1750]

[16] J.R. Flemish, K. Xie, J. Zhao [*Appl. Phys. Lett. (USA)* vol.64 (1994) p.2315]

[17] J.R. Flemish, K. Xie [*J. Electrochem. Soc. (USA)* vol.143 (1996) p.2620]

[18] J. Hong et al [*J. Electron. Mater. (USA)* vol.28 (1999) p.196]

[19] F. Lanois, P. Lassagne, D. Planson, M.L. Locatelli [*Appl. Phys. Lett. (USA)* vol.69 (1996) p.236]

[20] J.J. Wang et al [*Solid-State Electron. (UK)* vol.42 (1998) p.743]

[21] J.J. Wang et al [*J. Vac. Sci. Technol. A (USA)* vol.16 (1998) p.2204]

[22] P. Leerungnawarat et al [*J. Vac. Sci. Technol. B (USA)* vol.17 (1999) p.2050]

[23] F.A. Khan, I. Adesida [*Appl. Phys. Lett. (USA)* vol.75 (1999) p.2268]

[24] P. Chabert, N. Proust, J. Perrin, R.W. Boswell [*Appl. Phys. Lett. (USA)* vol.76 (2000) p.2310]

Chapter 5

Thermally grown and deposited dielectrics on SiC

E.Ö. Sveinbjörnsson and C.-M. Zetterling

5.1 CHAPTER SCOPE

This chapter gives a survey of the properties of dielectrics on SiC. Dielectrics are needed for surface passivation of SiC devices as well as a gate material in MOSFETs (metal-oxide-semiconductor field-effect transistors) and related structures for high-power and high-temperature operation. The natural dielectric of choice is silicon dioxide, which can be formed by a simple oxidation of the SiC. Despite significant progress in recent years, insufficient quality of the dielectric is of major concern. Below, we detail the methods currently used to fabricate dielectrics on SiC and discuss their applicability.

Because of the high electric breakdown strength of SiC, MOSFET transistors made in SiC are, in theory, expected to be superior to Si-based devices in terms of a very low on-resistance and high power capacity. However, in practice the SiC MOSFETs exhibit unacceptably low inversion layer mobilities, typically of the order of $50 \, cm^2/V \, s$, compared to bulk mobility of 500–$1000 \, cm^2/V \, s$. This is believed to be due to high interface state densities at the SiC/SiO$_2$ interface, especially near the conduction band edge. In addition, the MOSFETs are often unstable and exhibit threshold shifts upon use due to generation/annihilation of the defects at the SiO$_2$/SiC interface or charge trapping in the oxide.

5.2 THERMAL GROWTH OF SiO$_2$ AND DEPOSITED DIELECTRICS

5.2.1 Introduction

This section describes thermal oxidation of SiC as well as deposition of dielectrics. After discussing precleaning methods we detail the oxidation mechanism of SiC. Thereafter follows a description of the characterization methods used to evaluate the dielectrics and we demonstrate how the techniques are used for optimization of the dielectric.

5.2.2 Tutorial

5.2.2.1 Surface cleaning

Precleaning of the SiC wafer prior to oxidation has not been studied in great detail. The first step is degreasing using a sequence of organic solvents, for example trichlorethane, followed by acetone and ethanol. In wet cleaning, two approaches are mainly used: Piranha clean or RCA clean, followed by a dip in an HF-based solution to remove the native oxide [1,2]. Dry cleaning methods have also been investigated using, for example, hydrogen or ozone-plasma treatment before oxidation. In summary, there are indications that ozone plasma is beneficial, but the differences between the different studies of chemical preparation methods so far suggest that the key problem is not solved by altering the precleaning method. It is, however, possible that the crucial precleaning step is still to be discovered.

One way of somewhat improving the SiO_2/SiC interface is by the use of sacrificial oxidation. This is the process of oxidizing the SiC wafer, and removing the oxide grown with HF one or more times. Such methods were used in the early days of silicon processing, before the material growth was more or less fully mastered. The surface roughness of SiC epitaxial layers is also partially due to the off-axis growth that is presently needed to grow good-quality single-crystalline epilayers.

5.2.2.2 Thermal growth of SiO_2

As in the case of silicon, it is possible to oxidize SiC and form SiO_2 using wet (H_2O) or dry (O_2) oxidation in a diffusion furnace. In the case of silicon oxidation, the reaction is:

$$Si + O_2 \leftrightarrow SiO_2 \qquad (1)$$

The oxidation rate of SiC is much lower than that of Si and temperatures above 1000 °C are needed for reasonable processing times for oxidation. Details of the oxidation mechanism are unclear. The proposed reaction at the interface is between the silicon atom in the SiC lattice and the oxygen molecule

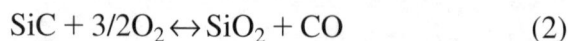

$$SiC + 3/2O_2 \leftrightarrow SiO_2 + CO \qquad (2)$$

An additional possible reaction is:

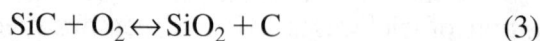

$$SiC + O_2 \leftrightarrow SiO_2 + C \qquad (3)$$

One of the concerns is the fate of the excess carbon. Most of it appears to escape through formation of CO gas, which is transported to the surface. But there is some evidence that carbon forms clusters at the SiO_2/SiC interface, which seriously affect the electrical properties of the interface [3]. There are some experiments and theoretical calculations that indicate differences in the seed suboxide at the interface that support this hypothesis [4,5]. One way of reducing the problem of excess carbon is to avoid consumption of the SiC by, for example, depositing the oxide. This has, with one exception, not been highly successful, as discussed in Section 5.2.2.3.

The oxidation kinetics are reasonably well described by the empirical model of Deal and Grove [6] for oxidation of silicon given by

$$t = \frac{X_0^2}{B} + \frac{X_0}{(B/A)} - \tau \tag{4}$$

where X_0 is the oxide thickness and t is the oxidation time. The time constant τ takes into account the initial oxide layer and can be neglected for SiC [1]. For long oxidation times the oxidation rate is reduced and the equation becomes:

$$X_0^2 = Bt \tag{5}$$

The oxidation is then limited by diffusion of the reacting species through the oxide. In the case of dry oxidation, the reaction is then limited by transport of oxygen through the silicon dioxide layer. More oxygen is consumed when oxidizing SiC as compared to Si (see EQNS (1) and (2)), which means that the diffusion-limited oxidation rate of SiC is 2/3 that of silicon. The parameter B is proportional to the diffusion coefficient of the reacting species and is called the parabolic rate constant. For short oxidation times EQN (4) reduces to:

$$X_0 = \frac{B}{A}(t + \tau) \tag{6}$$

In this case, the oxidation rate is limited by the reaction rate at the oxide/semiconductor interface. The constant B/A denotes the linear rate constant. The rate constants are thermally activated, i.e. they are proportional to $C \times \exp[-E_A/kT]$, where C is a constant, k is the Boltzmann constant, E_A the activation energy and T the oxidation temperature.

TABLE 5.1 gives a summary of literature data, with more details given in [7]. There is a large scatter in the data but it is clear that the C face oxidizes about three to ten times faster than the Si face, and dry oxidation is slower than wet oxidation, as is the case for oxidation of silicon. The oxidation rate differs for different faces of the SiC. For different crystal planes the rate is, in general, between the oxidation rates of the Si face and the C face. In addition, the oxygen rate is usually enhanced on ion-implanted regions, while the rate is sometimes suppressed in regions that have been dry etched.

For dry oxidation of the C-face the growth kinetics follow the linear-parabolic law and the activation energy E_B ($29\,kcal\,mol^{-1}$) agrees with the activation energy needed for an O_2 molecule to diffuse through the oxide. However, there are some discrepancies between the activation energies of the oxidation rate and the reacting species involved in the oxidation process. For example, the activation energy for diffusion of water in SiO_2 is $18\,kcal\,mole^{-1}$, which is significantly lower than the activation energy of the rate constant B in TABLE 5.1. This indicates that the oxidation rate is not determined only by diffusion of water through the oxide.

TABLE 5.1   Examples of experimentally determined activation energies of the linear and parabolic rates of oxidation growth in dry and wet oxygen for the 4H- and 6H-SiC polytypes. For details see [7] and references therein.

Dry oxidation			
Material/face	Temperature (°C)	$E_{B/A}$ (kcal mol^{-1})	E_B (kcal mol^{-1})
6H-C	1320–1420	25 ± 1	29 ± 1
6H-C	1050–1260	35	29
6H-Si	1050–1200	22.8	–
6H-Si	1320–1420	47 ± 14	141 ± 32
6H-Si	1050–1260	≈ 30	≈ 70
4H-Si	1050–1200	24.0	–
4H-Si	1200–1500	–	53–71
4H-C	1200–1350	–	28.7
4H-C	1350–1500	–	66.1
Wet oxidation			
6H-Si	1100–1250	65.7	127.4
6H-Si	1320–1420	–	106 ± 32
6H-C	850–1100	26	48
6H-C	1220–1420	37 ± 15	66 ± 5

As can be seen from the large scatter in data in TABLE 5.1, it is clear that test runs are needed to determine the actual oxide thickness obtained for a given diffusion furnace. FIGURE 5.1 shows an example of dry oxidation at $1200\,^{\circ}C$ as a function of oxidation time [8].

From the density of SiC the amount consumed during thermal oxidation can be calculated to be 46%, which is close to the value for thermal oxidation of Si. This is because the density of Si atoms in SiC is almost the same as in Si. To grow 100 nm of SiO_2 on SiC, 46 nm of SiC are consumed.

The optimal oxidation procedure itself has been a matter of debate and has, in some cases, been mystified without justification. The common procedure in accordance with the results of Shenoy et al [2] is to load the wafers at 700–$800\,^{\circ}C$ under oxygen flow and then ramp up the furnace (5–$10\,^{\circ}C\,min^{-1}$) to the oxidation temperature, which is typically between $1100\,^{\circ}C$ and $1200\,^{\circ}C$. The dry oxidation is done in pure O_2. The wet oxidation is done either by using water vapour from nitrogen or other carrier gas flowing through boiling deionized water, or using a pyrogenic steam (mixing hydrogen and oxygen and producing water vapour using a burner). The pyrogenic steam is a cleaner method and should be the standard method, even though some groups insist on using the water bubbler.

The temperatures needed for oxidation of SiC are higher than those needed for oxidation of silicon but the procedure is nevertheless compatible with standard silicon-based fabrication processing. It has therefore been natural to apply standard techniques of silicon oxidation directly on SiC.

This has not been very successful, and the SiC/SiO_2 interface, in general, has a density of interface state defects (low $10^{11}\,cm^{-2}eV^{-1}$ range) at least two orders of magnitude higher than the Si/SiO_2 interface. A rule of thumb in silicon devices is that, for successful operation, the mean interface state density should be in the $10^{10}\,cm^{-2}eV^{-1}$ range even though this is dependent upon the particular application. In the case of SiC-based devices, such a rule has not been established, but there is no reason to expect less stringent requirements on SiC devices. There seems to be no fundamental difference between the oxide quality on p- and n-type SiC. The difference in the effective fixed oxide charge appears to be mainly due to deep-lying donor interface states below midgap. These states are filled at room temperature under all circumstances in n-type material and do therefore remain undetected. In p-type SiC, however, these traps reveal their presence by being neutral or positively charged depending on the Fermi level. The oxide grown on the Si face is found to be superior in terms of electrical quality to that grown

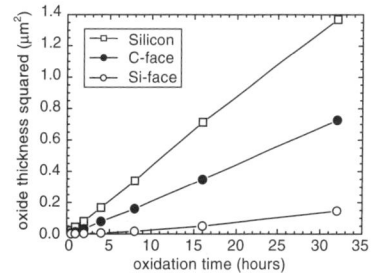

FIGURE 5.1 Dry oxidation of the silicon face and the carbon face of 6H SiC compared to Si at $1200\,^{\circ}C$.

on the C face and the Si face is therefore the preferred crystal face for applications.

5.2.2.3 Deposition of dielectrics

Several approaches have been used to reduce the interface state density and the fixed oxide charge. Many researchers suspect excess carbon at the interface to be the problem. One way of reducing this problem is to avoid consumption of the SiC by, for example, depositing the oxide. There are some scattered reports of this being successful. One of the methods is remote plasma-enhanced chemical vapour deposition (RPECVD). The main advantage of such a procedure is the possibility to do in situ plasma cleaning of the SiC surface prior to deposition, thereby obtaining better control over the SiC surface chemistry than during conventional thermal oxidation. This approach has been demonstrated by Gölz et al [9] but the growth method is still at a laboratory stage and it is highly uncertain if it ever will come into industrial use.

An additional method to avoid oxidation of SiC itself is using polysilicon deposition on top of the SiC. This is then followed by oxidation, which is terminated when all the polysilicon has been consumed and converted into silicon dioxide. Such an oxidation method is advantageous when fabricating UMOSFET where the dielectric covers different crystal planes of the SiC. If the gate oxide of a UMOSFET is thermally grown the oxide thickness varies with the crystal planes. By low-temperature oxidation of polysilicon it is possible to obtain uniform oxide thickness across the U-shaped gate. So far, oxides made by oxidation of polysilicon are not of the same quality as the native thermal oxide of SiC.

The most promising deposited oxides are grown with a specially designed process at room temperature called jet vapour deposition (JVD) [10,11]. The reacting gases flow towards the substrate through a nozzle, which provides high gas velocity when the species hit the semiconductor substrate. Using this technique, highly reliable dielectrics have been deposited on SiC, both oxides and oxide-nitride-oxide (ONO) structures. For example, the projected lifetime of n-channel MISFETs (metal–insulator–semiconductor field-effect transistors) using an ONO dielectric was 10 years at 450 °C at an electric field of 3.6 MV/cm within the dielectric stack. The mobility in fully processed MISFETs was, however, low at about 25 cm^2/V s at 450 °C so more experiments are needed to achieve acceptable mobility as well as reliability. Since the reason for the success of this deposition technique is not well established one might expect similar results to be obtainable using other more common types of plasma chemical vapour deposition.

There has been some interest in using alternative dielectrics. There are two main reasons for this, one being that silicon dioxide films are thus far not a real success on SiC in terms of inversion mobility and long-term reliability. Secondly, one would prefer a dielectric with a larger dielectric constant than silicon dioxide, due to the high-voltage applications of silicon carbide. Because of the high electric field within the silicon carbide, the dielectric has to withstand electric fields scaled by the ratio of the dielectric constants: see Chapter 7.

Silicon nitride has been investigated, but it has been found that the interface between silicon nitride and SiC is of too low a quality, as is the case for silicon. The nitride is also sensitive to carrier injection, and a proper interfacial barrier is needed to use nitride. One of the structures that has been investigated for high-voltage applications is a three-layer dielectric stack of oxide-silicon nitride-oxide. Such dielectrics have found use in the silicon industry. The purpose of the sandwiched nitride film is to increase the net dielectric constant of the stack. The oxide layers prevent carrier injection into the nitride. ONO structures have shown high breakdown fields, but the problem of low electron channel mobility remains.

Aluminium nitride is one of the dielectric candidates. One advantage is the possibility of growing AlN epitaxially on SiC. Even though high-quality epitaxial films of AlN in terms of morphology have been grown on SiC the insulating properties have thus far not been sufficient. There might be a fundamental reason for this since the bandgap of AlN is only about 6 eV. The band offsets between the valence and conduction bands of SiC and AlN are not known with certainty but there is evidence for a very low barrier between the valence bands (<0.5 eV). In any case, for 4H and 6H SiC we can in best cases expect 1.5 eV band offsets between the valence and conduction bands. This is presumably insufficient for high-voltage applications.

5.2.2.4 *Annealing of dielectrics*

Post-oxidation recipes are as numerous as are research groups. We will discuss a special case called re-oxidation a little later in the chapter but apart from that most groups perform nitrogen annealing (in some cases argon) for one hour at the oxidation temperature after oxidation. This step is believed to aid excess carbon to leave the oxide, but there is no direct experimental evidence for such an outdiffusion process. For example, Shenoy et al [2] found argon post-oxidation annealing to be necessary to prevent breakdown or degradation of the oxide during high-temperature characterisation (300–350 °C). After the post-oxidation anneal (if any) the furnace temperature is lowered, typically using a ramp of 5 °C min^{-1} to 700–800 °C before unloading. Some groups even lower the furnace temperature to room temperature before unloading, but there is no obvious advantage of using such a treatment.

Initially, the main effort was to reduce the density of deep-lying interface states, which for the most part act as effective fixed charges within the structure. A successful method for reducing such deep interface states is the so-called re-oxidation [12]. The physical reasons for its success are still not known. The key step of this method is annealing performed at approximately 950 °C in a wet ambient (H$_2$O vapour) typically for three hours. The name re-oxidation is perhaps somewhat misleading since there is negligible oxide growth at this temperature and this process is performed after formation of the oxide layer. Surprisingly, it appears as if the oxidation process itself is less crucial; both dry and wet oxides turn out to be state-of-the-art as long as the last step is this annealing procedure. Such oxides exhibit breakdown fields between 10 and 11 MV/cm. The electron channel mobility in MOSFETs made by such oxidation is typically between 10 and 50 cm^2/V s [12].

In the case of silicon, one of the main techniques to reduce the interface state density is hydrogen passivation of dangling bonds. This is usually done by so-called post-metallization annealing (PMA) at temperatures between 350 and 450 °C. When this process is successful it can reduce the interface state density down to values as low as 10^9 cm^{-2} eV^{-1}. In the case of SiC, however, no significant hydrogen passivation of interface states has been observed [3]. There are reports of reduction of deep-lying interface states after hydrogen annealing at high temperatures with an optimum temperature of about 800 °C. However, the interface state densities close to the band edges are not affected much by such an annealing. It is also possible that the high-temperature annealing treatment has an effect on thermal oxides, similar to the reoxidation process.

5.2.2.5 *Material characterisation*

A key parameter of interest is the thickness of the dielectric film, which is frequently estimated using light reflection or ellipsometry. In the light reflection technique monochromatic light is incident on the sample surface. At a certain angle some light will be directly reflected. However, light will also be reflected from the interface between the dielectric film and the substrate. For some wavelengths these waves will be in phase, while for other wavelengths destructive interference will occur. This enables an estimate of the film thickness to be made provided that the index of refraction is given. This method can be used down to a thickness of approximately 10 nm [13,14].

A more accurate technique is ellipsometry. The ellipsometry technique works in a similar manner to the reflectance technique but the incident light is polarized and the change in the polarization upon reflection from the dielectric/substrate interface is measured. This method can in best cases be applicable down to a dielectric thickness of 1–2 nm, but typically the data are only reliable down to about 10 nm [13,14].

The third technique used to estimate the dielectric thickness is capacitance measurement on an MOS test structure. This technique is discussed below.

5.2.2.6 Electrical characterisation

The key parameters used to determine oxide quality are leakage current, density of interface states, effective oxide charge, maximum breakdown field of the oxide and last, but not least, the carrier channel mobility in fully processed MOSFETs. Another important issue is the long-term reliability of the oxide. The characterisation techniques are essentially the same as those used for silicon-based devices although some precaution is needed to evaluate the data due to the wide bandgap of SiC. Here we concentrate on the use of capacitance–voltage analysis [13].

The standard method of analysing the electrical properties of oxides made on silicon is to combine high-frequency (100 kHz–1 MHz) and low-frequency or quasistatic capacitance–voltage (CV) measurements on MOS capacitors. Such analysis provides information on dielectric film thickness, interface state densities and the effective fixed oxide charge. In SiC-based structures the measurement procedure is nontrivial due to the large bandgap of SiC, and the quasistatic analysis at room temperature is not valid since interface states near the midgap do not emit their carrier charge within a reasonable time of measurement. The so-called effective fixed charge is indeed a combination of fixed charge in the oxide and immobile charge trapped in deep-lying interface states. It is necessary to heat the sample to about 350 °C to be able to fully mobilize this trapped charge [15]. This is rarely done in practice since most oxides degrade rather quickly at this temperature. A quick rough estimate of the oxide quality is therefore often made by using only the high-frequency curve at room temperature.

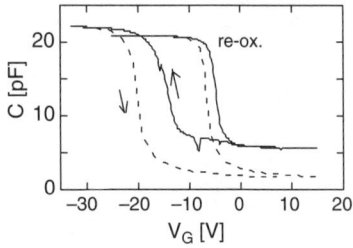

FIGURE 5.2 High-frequency CV data (1 MHz) of p-type 6H SiC MOS capacitors demonstrating the effect of re-oxidation (right set of curves) after dry oxidation (left set of curves). Dashed lines show the sweep from accumulation (-30 V) to depletion (15 V), full lines show the reverse sweep.

FIGURE 5.2 shows the effect of re-oxidation after dry oxidation on p-type 6H-SiC as visualised by high-frequency (1 MHz) CV data on MOS capacitors. A similar effect is observed on p-type 4H-SiC MOS structures. The oxide thickness is extracted from the capacitance in accumulation at -30 V, which equals the oxide capacitance $C_{ox} = A\varepsilon_r\varepsilon_0/d$, where A is the capacitor area, $\varepsilon_r\varepsilon_0$ is the dielectric constant of the oxide and d is its thickness.

The sample is first kept in accumulation (≈ -30 V) in darkness for about a minute, then the voltage is swept rapidly (≈ 1 V/s) towards deep depletion. The waiting time in accumulation is to ensure a complete recombination of any minority carriers. The fast sweep rate towards depletion is to minimize carrier emission from deep interface states during the sweep. Most of the interface states are therefore occupied by a majority carrier (hole) at the end of the first sweep. The sample is then illuminated with focused white light for one minute, and a majority of the interface states are emptied during this illumination period. The light is then turned off and shortly thereafter the bias is swept slowly back towards accumulation. During the backwards sweep, two processes occur. First, inversion charge is removed from the interface. In the second process, interface states recapture holes from the valence band and the capacitance in accumulation approaches the initial value. The hysteresis observed is therefore a rough measure of the number of interface states in the sample. The interface state density in the re-oxidized sample is about a quarter of the density in the dry oxide sample.

The theoretical flatband voltage, i.e. the gate voltage corresponding to no band bending in the semiconductor, for the structure in FIGURE 5.2 is -2.6 V. The flatband voltage for the oxidized sample (no light) is about -20 V, while the flatband voltage of the reoxidized sample is about -5 V. The difference between the theoretically and experimentally determined flatband voltages is a measure of the effective oxide charge, which is a sum of the fixed oxide charge and immobile charge trapped in deep interface states. It is, however, difficult to differentiate between these two contributions from these measurements. As mentioned earlier, one needs to perform the CV analysis at elevated temperatures (350–400 °C) to enable this separation.

Recently, attention has been focused on characterising n-type 4H-SiC MOS capacitors since it is recognized that the most probable cause of low electron channel mobility in MOSFETs is trapping of electrons in shallow interface states near the conduction band of 4H-SiC.

Such states are, for example, detected in high-frequency CV analysis as depicted in FIGURE 5.3. The picture shows

CV curves of an n-type 4H MOS capacitor for three different measurement frequencies. The sample bias is swept from accumulation (positive voltages) towards depletion (negative voltages). The flatband capacitance is 19 pF. We notice the frequency dispersion of the curves near flatband, which is dispersion due to interface states that are partially responding to the AC signal. The electron emission rate of these states is within the frequency range of the probing signal. This means that at high frequency (100 kHz) a smaller number of interface states respond to the probe signal and such interface states are filled with electrons, giving rise to a shift (or stretching) of the curve towards positive voltages. When the probe frequency is lowered, a larger portion of the interface states is able to follow the probe frequency; less charge is trapped and the CV curve is shifted less towards positive voltages. This frequency dispersion is a fingerprint of interface states near the conduction band edge and is observed in virtually all oxides grown on n-type 4H-SiC. A similar dispersion in high-quality oxides on n-type 6H-SiC is negligible in comparison. This frequency dispersion is used to estimate the density of interface states as a function of energy in the SiC bandgap near the conduction band edge [16]. Of all the investigations so far reported it seems that annealing of the oxide in NO or N_2O ambient results in a significant reduction of these interface states. We discuss this in more detail in Section 5.3.

The conductance method is an additional technique [13] that allows an estimate of the density of interface states. It is based on the measurement of the equivalent parallel conductance of an MOS structure as a function of bias and frequency. The method is time-consuming when compared to CV analysis. This technique is considered to be the most sensitive method to determine interface state densities, and trap densities of $10^9\,cm^{-2}\,eV^{-1}$ and lower can be measured in silicon. In the case of SiC the method is more problematic. A major source of error is the surface potential fluctuations, which smear out the signal and distort the conductance peak one is searching for. As in the CV analysis it is necessary to heat the sample to about 350 °C to obtain information on density of interface states near midgap. At room temperature the technique monitors interface states in a rather narrow energy range within the SiC bandgap typically between 0.2 and 0.5 eV from the majority carrier band edge.

Apart from the CV and conductance methods, techniques like deep level transient spectroscopy (DLTS) and charge pumping have been used to a limited extent. The charge-pumping method monitors charge emission from interface states in MOSFETs. The method is highly sensitive to gate leakage current and the

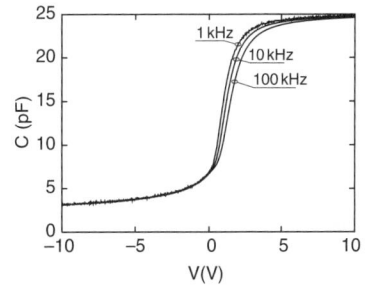

FIGURE 5.3 High-frequency CV data of a dry-oxidized n-type 4H SiC MOS capacitor showing the frequency dispersion due to interface states near the SiC conduction band edge.

quality of current SiC-based MOSFETs makes this method infeasible for common use. DLTS is used to estimate the density of interface states as a function of energy within the bandgap. The method is cumbersome on SiC and indeed on MOS structures in general, and is therefore not frequently used. A simpler (related) technique is the thermally stimulated current (TSC) analysis. TSC is based on monitoring the displacement current due to charge emission from interface and/or bulk traps in the semiconductor. Such studies of n-type 4H-SiC MOS capacitors reveal that there is a peak in density of interface states located approximately 0.15 eV below the conduction edge of 4H-SiC [17].

5.2.3 Applicability

Dielectrics are needed for surface passivation and as gate dielectrics in field-effect transistors aimed primarily at high-power, high-frequency and high-temperature applications. Most research groups are investigating dielectrics for gate insulation but the requirements on the dielectric for surface passivation are similar.

Planar structures are difficult to fabricate because of the extremely low diffusion coefficients of dopants in SiC. Ion implantation of SiC is difficult as well due to extensive lattice damage during the ion implantation, which is not healed unless the SiC is annealed afterwards at very high temperatures, typically 1200–1700 °C depending on the ion-implantation species. These high annealing temperatures degrade the gate dielectric, which prevents the use of self-aligned polysilicon gates. The ion implantation and the subsequent annealing are therefore done prior to the formation of the gate dielectric.

The problem of lattice damage associated with ion implantation is one of the reasons that vertical power device structures have received much attention, for example U-shaped grooved MOS. In such structures the source and drain regions consist of epitaxially grown material and ion implantation is avoided. However, one major complication in such vertical power devices is local breakdown of the gate oxide at trench corners. Another problem concerns the vertical sidewalls of such structures, since the oxide growth rate and quality depend substantially upon the surface orientation and whether the surface is silicon-face or carbon-face. The UMOSFET structures are therefore preferably made using deposited dielectrics or by oxidation of polysilion, as previously mentioned. Such devices have been demonstrated with an electron channel mobility comparable to that of lateral devices [18].

5.2.4 Comparison of techniques

In general, thermal oxides are of better quality than deposited oxides. Experimental observations suggest that dry oxides have better reliability than wet oxides (this is the case for oxides made on Si as well). On the contrary, wet oxides have a lower density of deep-lying interface states and less effective oxide charge. However, dry oxides that receive a re-oxidation treatment (950 °C, 3 h) in wet ambients exhibit similar interface state densities to wet oxides.

Lipkin and Palmour [12] made an extensive study comparing several deposited dielectric stacks and thermal oxides. They found that in terms of reliability oxide-nitride-oxide (ONO) structures were superior to thermal and deposited oxides. Highly reliable dielectrics have also been reported using jet vapour deposition (JVD). For example, the projected lifetime of n-channel MISFETs using an ONO dielectric was 10 years at 450 °C at an electric field of 3.6 MV/cm within the dielectric stack [11].

The highest electron channel mobility reported in MOSFETs to date is 176 cm^2/V s in 4H-SiC and 113 cm^2/V s in 6H-SiC [19]. These values are about twice the best values reported by other authors. However, a part of the explanation is that these estimates were done at unusually low semiconductor surface fields (1×10^5 V/cm), and the mobility decreases considerably with increasing electric field. More extensive work is needed to seek the correlation between the channel mobility, fixed oxide charge, interface trapped charge and oxide breakdown field. It is commonly believed that the reason for low mobility is surface scattering and in particular trapping of carriers in shallow interface states located within the oxide adjacent to the interface. This problem is more severe in 4H-SiC than in 6H-SiC [20–22]. The origin of near interfacial defects, often called border traps, is as yet unknown, but they are generally believed to be responsible for the low channel mobilities observed in SiC-based MOSFETs. The same types of defects have been detected by internal photoemission studies in SiO$_2$ grown on various polytypes of SiC as well as on Si. Due to differences in bandgap, however, these defects are inactive in Si MOS structures [3].

The main parameter determining the electron channel mobility in n-channel MOSFETs appears to be the density of the above-mentioned interface states near the conduction band edge. It is, however, difficult to compare the density of such interface states for differently prepared oxides from various research groups. The extracted interface state density increases exponentially towards the conduction band edge, which makes it practically impossible to compare values from different authors. Instead, we summarize

TABLE 5.2   Field effect electron mobility in n-channel MOSFETs reported in the literature.

Polytype	Crystal face	Dielectric	μ_{FE} (cm^2/V s)	Ref.
6H	Si	wet oxidation 1025 °C + re-ox. 950 °C	40–50	[12]
6H	Si	ONO (thermal oxide-deposited nitride-oxidized nitride)	72	[12]
6H	Si	thermal oxide 1100 °C	24	[21]
6H	Si	low-temperature oxide 8000 Å, followed by wet oxidation 1100 °C + re-ox. 950 °C	110	[19]
6H	Si	wet oxidation	45	[23]
6H	11–20	wet oxidation	36–116	[23]
15R		thermal oxide 1100 °C	33	[21]
15R		dry oxide	47	[24]
15R		wet oxide	50	[24]
4H	Si	low-temperature oxide 8000 Å, followed by wet oxidation 1100 °C + re-ox. 950 °C	160	[19]
4H	Si	dry oxide	5	[25]
4H	Si	dry oxide + NO passivation	30	[25]
4H	Si	wet oxide	5	[25]
4H	Si	wet oxide + NO passivation	37	[25]
4H	Si	thermal oxide partially grown in NO	48	[26]
4H	Si	dry oxide 1200 °C	5–10	[27]
4H	Si	dry 1200 °C + re-ox. 950 °C	47	[27]
4H	11–20	wet oxidation 1150 °C	32	[28]
4H	11–20	wet oxidation 1150 °C + H$_2$ anneal 800 °C	110	[28]
4H	Si	wet oxidation	5.6	[23]
4H	11–20	wet oxidation	82–96	[23]
4H	03–38	wet oxidation	10	[29]

literature data in TABLE 5.2 using the reported peak field effect electron mobility in n-channel MOSFETs as a parameter.

5.2.5   Overcoming problems

The use of dielectrics as gate material in SiC MOSFETs is hampered by the low electron channel mobility observed. This key problem prevents commercialisation of any such devices. The origin of the problem is known; the channel electrons in n-channel devices are trapped in interface states located close to the SiC conduction band edge. The physical origin of these states is as yet unknown, but they must be removed or their density reduced to acceptable levels. These states appear in thermally grown as well as deposited oxides. At present it appears as if annealing the dielectric in NO ambient lowers the density of these interface states but the channel mobility reported in such devices is still too low for practical use. The idea of replacing the silicon dioxide with another dielectric has thus far not been successful but the use of dielectric stacks such as oxide-nitride-oxide (ONO) improves the reliability of the devices.

5.3 RECENT DEVELOPMENTS AND FUTURE TRENDS

The low electron mobility in MOSFETs fabricated on 4H-SiC is of major concern and researchers are constantly looking for new ways of growing a dielectric. One promising method that has recently been used is annealing in NO or N_2O ambients.

5.3.1 Oxide growth and annealing in NO and N_2O ambients

There are still few reports on this matter, to some extent due to the highly toxic nature of NO, which prevents it from being used in most laboratories. It is clear that post-oxidation annealing in NO has an effect on the SiO_2/SiC interface and there are indications that this results in an improvement of channel mobility in fully processed MOSFET devices [25,26]. However, the actual physical process involved is unclear, but more efficient removal of carbon (by, for example, CN products) has been suggested. Evidently, the annealing in NO ambient results in a pile-up of nitrogen at the oxide/SiC interface and this results in a reduction of interface states near the conduction band edge of 4H-SiC [16,30,31]. A similar reduction is achieved using NH_3 or N_2O ambient, although annealing in ammonia causes premature oxide breakdown. The annealing temperatures that give beneficial properties are typically 1200–1300 °C. There is a significant oxidation of the SiC at this temperature, so it is as yet unclear how the interface passivation takes place.

5.3.2 Dielectrics grown on the [11$\bar{2}$0] and the [0338] surfaces

Another new approach is to grow oxides on wafers oriented in the [11$\bar{2}$0] or the [0338] direction [23,29]. Wafers grown in the

[11$\bar{2}$0] direction are now commercially available on a small scale and the initial studies indicate that devices based on such structures exhibit higher channel mobility [23]. This is still at the laboratory stage and the advantages of using this crystal orientation are not yet clearly demonstrated.

5.4 CONCLUSIONS

The use of dielectrics in SiC processing is currently limited to passivation layers. The interfaces between the native oxide as well as the deposited oxide on SiC are not of sufficient quality for use as a gate dielectric in switching devices. Reliable oxides with breakdown properties comparable to those of oxides grown on silicon have been demonstrated. However, the electron channel mobility in SiC-based MOSFETs is very low (10–$50\,cm^2/V$ s) due to severe trapping at interface states. This problem is more serious in 4H-SiC than 6H-SiC. Methods to reduce the number of these detrimental interface defects are being developed and we do see progress. The question still remains as to whether this key problem can be solved to enable commercial gate-controlled devices using dielectrics.

REFERENCES

[1] J.J. Kopanski [in *Properties of Silicon Carbide*, EMIS Datareviews Series No. 13 (INSPEC, London, 1995) p.121–9]

[2] J.N. Shenoy, et al [*J. Electron Mater. (USA)* vol.24 (1995) p.303]

[3] V.V. Afanas'ev, M. Bassler, G. Pensl, M. Schulz [*Phys. Status Solidi A (Germany)* vol.162 (1997) p.321 and references therein]

[4] C. Virojanadara, L.I. Johansson [*Surf. Sci. Lett. (Netherlands)* vol.472 (2001) p.L145]

[5] R. Buczko, S.J. Pennycook, S.T. Pantelides [*Phys. Rev. Lett. (USA)* vol.84 (2000) p.943]

[6] B.E. Deal, A.S. Grove [*J. Appl. Phys. (USA)* vol.36 (1965) p.3770]

[7] C. Raynaud [*J. Non-Cryst. Solids (Netherlands)* vol.280 (2001) p.1 and references therein]

[8] C.-M. Zetterling [PhD thesis (KTH, Royal Institute of Technology, Stockholm, Sweden, 1997)]

[9] A. Gölz, G. Lucovsky, K. Koh, D. Wolfe, H. Niimi, H. Kurz [*Microelectron. Eng. (Netherlands)* vol.36 (1997) p.73 and references therein]

[10] X.W. Wang, Z.J. Luo, T.P. Ma [*IEEE Trans. Electron Devices (USA)* vol.47 (2000) p.458]

[11] W.J. Zhu, X.W. Wang, T.P. Ma, J.B. Tucker, M.P. Rao [*Mater. Sci. Forum (Switzerland)* vol.338–342 (2000) p.1311]

[12] L.A. Lipkin, J.W. Palmour [*IEEE Trans. Electron Devices (USA)* vol.46 (1999) p.525; *J. Electron. Mater. (USA)* vol.25 (1996) p.909]

[13] D.K. Schroder [*Semiconductor Material and Device Characterization* (Wiley, New York, USA, 1990)]

[14] J.D. Plummer, M.D. Deal, P. Griffin [*Silicon VLSI Technology* (Prentice Hall, Upper Saddle River, NJ, USA, 2000)]

[15] J.A. Cooper Jr. [*Phys. Status Solidi A (Germany)* vol.162 (1997) p.305]

[16] G.Y. Chung et al [*Appl. Phys. Lett. (USA)* vol.76 (2000) p.1713]

[17] H.Ö. Ólafsson et al [*Mater. Sci. Forum (Switzerland)* vol.389–393 (2002) p.1001]

[18] S. Scharnholz, O. Hellmund, J. Stein, B. Spangenberg, H. Kurz [*Mater. Sci. Forum (Switzerland)* vol.353–356 (2001) p.651]

[19] S. Sridevan, B.J. Baliga [*Mater. Sci. Forum (Switzerland)* vol.264–268 (1998) p.997]

[20] R. Schörner, P. Friedrichs, D. Peters [*IEEE Trans. Electron Devices (USA)* vol.46 (1999) p.533]

[21] R. Schörner, P. Friedrichs, D. Peters, D. Stephani [*IEEE Electron Device Lett. (USA)* vol.20 (1999) p.241]

[22] N.S. Saks, S.S. Mani, A.K. Agarwal [*Mater. Sci. Forum (Switzerland)* vol.353–356 (2001) p.1113; *Appl. Phys. Lett. (USA)* vol.76 (2000) p.2250]

[23] H. Yano, T. Hirao, T. Kimoto, H. Matsunami, K. Asano, Y. Sugawara [*Mater. Sci. Forum (Switzerland)* vol.338–342 (2000) p.1105]

[24] H. Yano, T. Kimoto, H. Matsunami, M. Bassler, G. Pensl [*Mater. Sci. Forum (Switzerland)* vol.338–342 (2000) p.1109]

[25] G.Y. Chung et al [*IEEE Electron Device Lett. (USA)* vol.22 (2001) p.176]

[26] R. Schörner, P. Friedrichs, D. Peter, D. Stephani, S. Dimitrijev, P. Jamet [*Appl. Phys. Lett. (USA)* vol.80 (2002) p.4253]

[27] R. Kosugi et al [*Mater. Sci. Forum (Switzerland)* vol.389–393 (2002) p.1049]

[28] J. Senzaki et al [*Mater. Sci. Forum (Switzerland)* vol.389–393 (2002) p.1061]

[29] T. Hirao, H. Yano, T. Kimoto, H. Matsunami, H. Shiomi [*Mater. Sci. Forum (Switzerland)* vol.389–393 (2002) p.1065]

[30] P. Jamet, S. Dimitrijev [*Appl. Phys. Lett. (USA)* vol.79 (2001) p.323]

[31] H. Li, S. Dimitrijev, D. Sweatman, H.B. Harrison, P. Tanner [*J. Appl. Phys. (USA)* vol.86 (1999) p.4316]

Chapter 6

Schottky and ohmic contacts to SiC

C.-M. Zetterling, S.-K. Lee and M. Östling

6.1 CHAPTER SCOPE

This chapter covers Schottky (rectifying barrier) and ohmic (non-rectifying, no barrier) contacts to SiC. There are a few devices such as the MESFET and the Schottky diode that actually need Schottky contacts. However, for most devices ohmic contacts are preferred between a metal and semiconductor. To understand the formation of ohmic contacts, the theory for Schottky contacts is needed, and it is therefore covered first in this chapter. The formation methods are similar for both contact types, and are covered mainly in the section on Schottky contacts. The only way to know if a contact is good enough is to measure its electrical characteristics, and therefore electrical characterisation methods are covered in detail for both contact types.

6.2 SCHOTTKY CONTACTS

6.2.1 Introduction

This section concerns Schottky contacts, and will start with some techniques for forming Schottky contacts, and material characterisation. The theory for thermionic emission is described next, and some different methods of measuring the barrier height. To form a Schottky contact, usually low doping is required, $10^{17}\,cm^{-3}$ and below.

6.2.2 Tutorial

6.2.2.1 Formation of the Schottky contact

There are four main steps to forming metal contacts on a semiconductor: cleaning of the surface, metal deposition, patterning of the metal, and annealing.

The purpose of the cleaning step is to make sure that the deposited metal can come into intimate contact with the SiC. Normal contaminants are usually easy to remove with different etchants, very similar to Si technology. If excess surface damage is suspected, either from extensive reactive ion etching or ion implantation, a sacrificial oxidation step can be employed prior to metal deposition. This sequence consists of thermally growing silicon dioxide, which consumes the damaged top layer of SiC, and then removing this oxide by wet etching. Probably the most difficult task is to keep the interface free from silicon dioxide. Native oxide will form even at room temperature using oxygen from the ambient, and even a final HF step prior to loading the samples cannot keep all oxide away. If the metal is to be deposited at high temperature in vacuum, a preclean step can be done in situ, just before metal deposition, but temperatures in excess of 900 °C are needed. In a sputter deposition equipment the surface can be cleaned using backsputtering. A common reason for leakage current or nonideality is an unclean interface. If a high-temperature clean cannot be performed, a careful choice of metal has to be made; see Section 6.2.4 below.

There are three common methods of metal deposition used in SiC device fabrication: sputtering, evaporation and CVD, in order of applicability. Sputtering is most commonly used today, since it is a fast and economical method. Ions are extracted from a plasma discharge and allowed to sputter metal from a target, and this metal is then deposited on the SiC sample. Metal adhesion is good, compound targets are possible, and the surface can be precleaned by backsputtering (similar to reactive ion etching). The glow discharge can either be inert species (argon) or reactive species (nitrogen). Evaporation is also common, since it allows higher deposition rates and can be performed in an ultra-clean vacuum system. Higher demands are usually placed on the vacuum system, and adhesion can be a problem. Multiple atom layers can be deposited, by either sequential evaporation, or co-evaporation using multiple sources. Evaporation sources are thermal (a high current is passed through a crucible containing the metal), electron beam (a high current electron beam is focused on the metal) or effusion sources. Metals can also be deposited by chemical vapour deposition (CVD), but this is not as common as the other two. One advantage with CVD methods is that epitaxial growth is easier to promote, because of the thermal energy present during the deposition. Some co-evaporation methods and reactive sputtering methods have also been successfully used to deposit epitaxial metal films.

The patterning of the metal can be done either with a lithographical method prior to metal deposition (lift-off), or afterwards through etching. The lift-off process relies on the deposited metal not covering the steps in the photoresist pattern, so that acetone in an ultrasonic bath can remove the photoresist and the metal deposited on top of it (FIGURE 6.1). The deposited metal layer must be thinner than the photoresist used in the lift-off. Preferably, a photoresist with sharp edges or a negative profile should be used, or a two-step photoresist, for a clean lift-off. Evaporation gives the best results since the source of metal is directional and the deposition process does not damage the resist. Sputtering may damage the photoresist and have too good a step coverage, resulting in remaining metal or ragged edges. To use lift-off with CVD, extra steps need to be performed to pattern silicon dioxide or another dielectric for the lift-off, since the photoresist cannot withstand the high temperatures of a CVD process. Patterning of the deposited metal afterwards is easier, if etch recipes exist that can wet or dry etch the metal without damaging the photoresist. Often, multiple layers are used in a metallisation stack, and in these cases lift-off may be preferred, even though the yield is usually worse compared to etching.

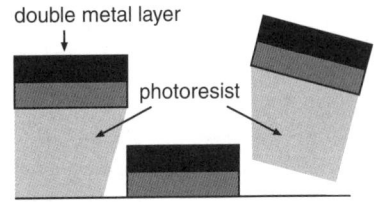

FIGURE 6.1 Patterning of multiple metal stacks using lift-off.

The final step of the contact formation is the annealing. Usually, an alloying between the metal and the SiC is unwanted, unlike the ohmic contact formation (see Section 6.3.2 below). For this reason the annealing is not done at too high a temperature, around 500 °C. The annealing step can improve the Schottky contact significantly, reducing leakage and improving the ideality factor. This is believed to be caused by the remaining silicon dioxide breaking up, and allowing intimate contact between the metal and the SiC, and by passivation of interface states. Titanium (Ti) is well known for promoting this process, since TiO_2 readily forms at these low temperatures. The annealing process is normally performed in vacuum, or in a nonreactive ambient such as Ar or N_2.

6.2.2.2 *Material characterisation of Schottky contacts*

If the contact process is not well established, some material characterisation can be used to investigate reactions between the metal and the SiC. These investigations are usually performed on blanket-deposited samples (unpatterned) so that a sufficiently large area is available for the characterisation method. Normally, the sample is analysed before and after the annealing process, to discover possible reactions. Using X-ray diffraction (XRD) some metal alloys can be identified. This is based on the fact that the formed alloys have crystalline or polycrystalline phases, which have lattice spacings that can be measured with XRD. Rutherford backscattering spectrometry (RBS) is used for identifying atoms by their mass, by measuring the energy of helium nuclei backscattered from the surface. If the metals used have large differences in atomic mass this method can give quite high resolution. Examples of the applicability of some of the material characterisation methods are shown in Section 6.3.2 for ohmic contacts, since it is easier to see any difference when a 950°C anneal has been performed.

6.2.2.3 Electrical characterisation of Schottky contacts

When a metal and a semiconductor are brought into contact, the Fermi levels in the two materials will line up at equilibrium. The Fermi level in the metal, also called the work function of the metal, does not usually line up with the Fermi level in the semiconductor, where it depends on the doping. This causes the bands to bend in the semiconductor, as in FIGURE 6.2, and a depletion layer forms under the contact. In forward bias (positive voltage on the metal for n-type, negative for p-type) a current can flow. In reverse bias very little current can flow, even taking into account image force lowering of the Schottky barrier, until the field is high enough that breakdown occurs through impact ionisation and avalanche multiplication. The detailed equations for the operation of Schottky contacts are derived in several books [1,2].

When Schottky contacts are characterised electrically, they are normally referred to as Schottky diodes. EQN (1) below describes the current density J for a Schottky diode.

$$J = J_S\left[\exp\left(q\frac{V - JR_{on}}{\eta kT}\right) - 1\right]$$

$$\text{where } J_S = A^*T^2\exp\left(-\frac{q\phi_b}{kT}\right) \tag{1}$$

Here, V is the applied voltage, η is the ideality factor, A^* is the effective Richardson's constant (146 A cm^{-2}K^{-2} for SiC) and ϕ_b is the effective Schottky barrier height. T is the absolute temperature in Kelvin. A typical current-voltage (IV) measurement of Schottky diodes is shown in FIGURE 6.3. For a large voltage range a linear curve is seen in the logarithmic current plot, which corresponds to the exponential voltage dependence, while V is much larger than JR_{on}. At higher currents the specific on-resistance will limit the current, since a large part of the applied voltage will be dropped over the low-doped region, and V is reduced by the current density J multiplied by the on-resistance. R_{on} is the specific on-resistance, which can be extracted from the linear portion of a linear plot of J versus V. For the linear region

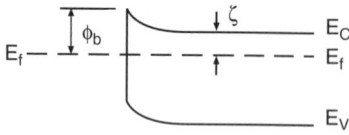

FIGURE 6.2 The formation of a Schottky barrier on n-type SiC.

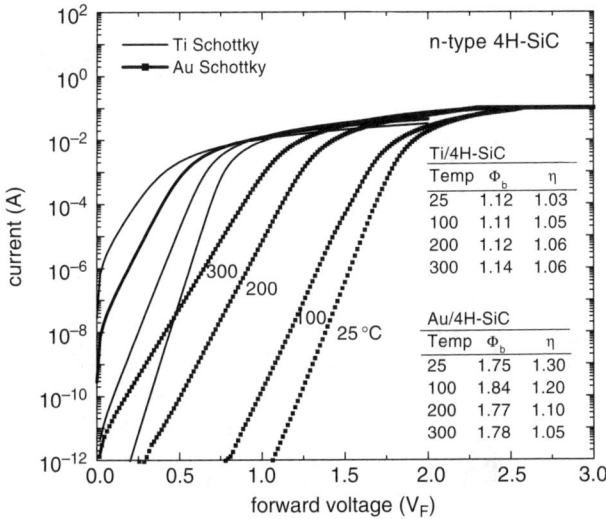

FIGURE 6.3 Forward current versus forward bias voltage for Ti and Au Schottky diodes on n-type SiC measured at several temperatures.

in the logarithmic J versus V plot, the Schottky barrier height and the ideality factor can be extracted from the y-intercept (J_S) and the slope ($dV/d\ln J$), respectively:

$$\phi_b = \frac{kT}{q}\ln\left(\frac{A^*T^2}{J_S}\right) \quad \text{and} \quad \eta = \frac{q}{kT}\left(\frac{\partial V}{\partial \ln J}\right) \qquad (2)$$

By doing the IV measurements and extraction at several temperatures, as in the figure, better values can be expected. Normally, the current I is measured, and the current density is calculated from the contact area A ($J = I/A$). The ideality factor cannot be smaller than around 1.01 theoretically [1], and good Schottky contacts normally have an ideality factor below 1.20.

For a certain metal, it would be expected that the sum of the Schottky barrier heights on n-type and p-type material would be equal to the energy bandgap of the semiconductor. This is approximately true for most metals on a well-cleaned SiC surface, and is referred to as the Schottky-Mott limit [1]. There may be an offset due to Schottky barrier pinning at the interface due to surface states, and if this completely determines the Schottky barrier it is referred to as the Bardeen limit [3]. In practice, the metal work function determines the Schottky barrier roughly, but the surface states may influence the value slightly. If we look at the measurements on p-type diodes (FIGURE 6.4) we can see another feature specific to SiC at high currents. Due to the incomplete ionisation of p-type dopants discussed in Chapter 3, the on-resistance of the p-type region is not constant, but is reduced as the temperature is increased.

115

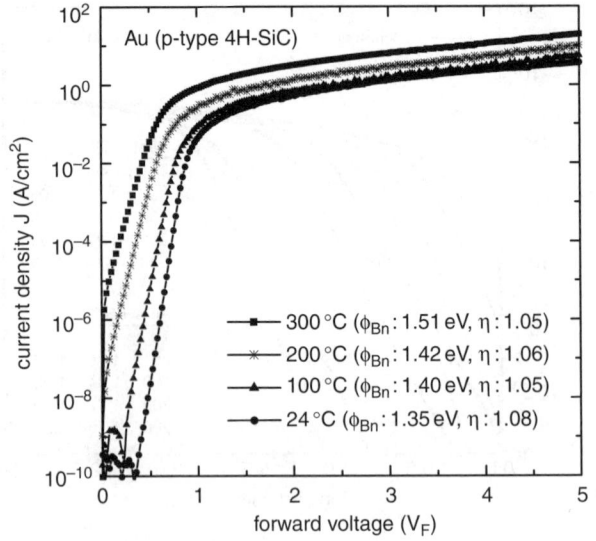

FIGURE 6.4   Current density versus forward bias voltage for Au Schottky diodes on p-type SiC measured at several temperatures.

In reverse bias very little current flows, but instead high-frequency capacitance-voltage measurements can be performed using a small AC voltage (50 mV) superimposed on the DC bias. Commercial equipment is available from several vendors, and typically measures the capacitance in the frequency range 100 kHz–1 MHz. The capacitance C of a reverse-biased Schottky diode is determined by the Schottky barrier height ϕ_b, the reverse voltage V_R and the donor concentration N_d:

$$C = A\left(\frac{qN_d\varepsilon_s}{2}\right)^{1/2}\left(\phi_b - \xi + V_R - \frac{kT}{q}\right)^{-1/2} \qquad (3)$$

where A is the area of the contact, and ξ is the difference in energy between the Fermi level and the conduction band in the bulk of the device (n-type). Squaring and inverting this relation we find

$$\frac{1}{C^2} = \left(\frac{2}{A^2qN_d\varepsilon_s}\right)\left(\phi_b - \xi + V_R - \frac{kT}{q}\right) \qquad (4)$$

$$N_d = \frac{2}{q\varepsilon_s\left(\frac{\partial(1/C^2)}{\partial V_R}\right)A^2} \qquad (5)$$

which implies that the Schottky barrier height can be determined from the x-axis intercept of a $1/C^2$ versus V_R plot. Moreover, the

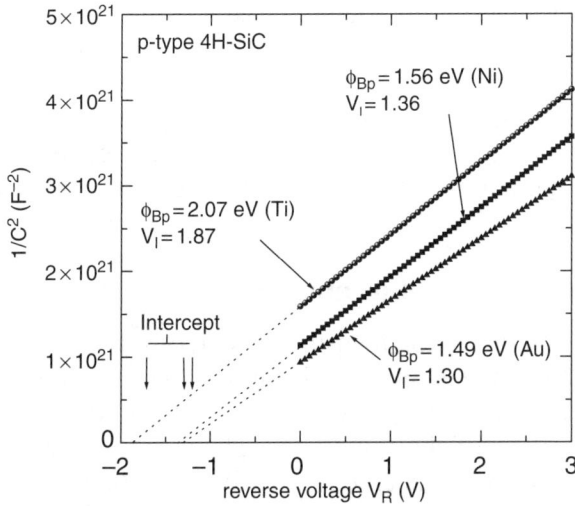

FIGURE 6.5 $1/C^2$ versus reverse bias plots for Ti, Ni and Au Schottky diodes to p-type 4H SiC measured at 100 kHz at room temperature. Diode area is 1.26×10^{-3} cm^2.

doping concentration can be determined from the inverse slope, and if the doping is constant, a straight line should result. Typical CV curves are shown in FIGURE 6.5.

A third method to determine the Schottky barrier height is photoelectric measurements. If monochromatic light is incident upon the metal surface, the resulting photocurrent can be measured versus wavelength of the light. For photon energies larger than the barrier energy by 3 kT, the square root of the photocurrent Y is linearly dependent on the photon energy, and the Schottky barrier height can be extracted from the x-axis intercept: see EQN (6). The disadvantage with this method is that a monochromatic light source with variable wavelength is needed.

$$Y \approx (h\nu - q\phi_b)^2 \quad \text{for } h\nu - q\phi_b > 3\,kT \qquad (6)$$

6.2.2.4   Applicability

The current transport across the Schottky barrier is strongly field dependent. EQN (1) above is mostly usable for forward bias, and will not yield correct results for the reverse current even if barrier lowering from the reverse voltage is taken into account [4]. However, there is a clear trend in the leakage current where more leakage is seen for smaller barriers. The reason for this is that since SiC can handle higher critical fields (see Chapter 1), the drift region is designed with quite high doping, which results in high electric fields. Also, there is electric-field enhancement at the edge of the contact metal region. Therefore, both field emission and thermionic field emission, which is tunnelling-type transport, will dominate over thermionic emission. These current mechanisms are further discussed in Section 6.3.2 below for ohmic contacts. The method to overcome the increased leakage from high electric fields is termination of the edge of the metal, and these junction-termination techniques are further discussed in Chapter 7.

6.2.3 Comparison of techniques

The three methods for measuring Schottky barrier height have different advantages and disadvantages. The advantage with IV measurements is that the ideality factor gives some indication of the quality of the Schottky diodes. The on-resistance can also be calculated from a linear current versus voltage plot for high currents. The disadvantage is that a good ohmic contact is needed to the SiC. CV measurements are less sensitive to the contact quality, and also have the advantage of allowing the doping concentration to be measured. If the doping concentration varies, the results can be difficult to interpret. Photoelectric measurements are seldom performed, but could be a last resort if the other methods are unable to show a result. There is not really a choice between IV and CV; normally both methods are performed since they are so easy. However, the measured Schottky barrier height is normally slightly higher for CV measurements, because local Schottky barrier lowering influences current measurements more (exponential dependence) than capacitance measurements (averaged over the contact area).

Most references in the literature are from measurements of the Schottky barrier height on 3C and 6H SiC, since these were the two polytypes first investigated. More recently, the 4H polytype has been preferred, and some selected data from the literature are presented in TABLE 6.1. Comparing 4H to 6H, the Schottky barrier heights on p-type are more or less the same, which is not surprising since the valence bands coincide in the SiC polytypes. The 0.2 eV difference in the bandgap energy between 4H and 6H SiC is mainly in the conduction bands, and therefore the Schottky barriers to n-type material are about 0.2 eV higher on 4H than on 6H SiC.

6.2.4 Overcoming problems

Many problems can occur in manufacturing the contacts, even though the basic process is simple. Adhesion problems (the metal peels off, or rolls up) are caused by poor vacuum and badly cleaned surfaces. The problems are greater if there is much mechanical stress in the deposited metal due to differences in

TABLE 6.1   Selected measurements of Schottky barrier heights
on 4H SiC from the literature.

n or p	Metals	Barrier height (eV)		η	Face	Comments	Ref.
		I-V	C-V				
n	Ti	0.80	–	1.15	Si-	as-deposited (20 °C)	[5]
		0.85	–	1.10		122 °C	
	Ni	1.30	–	1.21		20 °C	
		1.40	–	1.12		122 °C	
		1.50	–	1.12		255 °C	
n	Au	1.73	1.85	1.02	Si-		[6]
		1.80	2.10	–	C-		
	Ni	1.62	1.75	1.20	Si-		
		1.60	1.90	–	C-		
	Ti	0.95	1.17	–	Si-		
		1.16	1.30	–	C-		
n	TiW	1.22	1.23	1.05	Si-	as-deposited	[7]
		1.18	1.19	1.10		500 °C, 30 min	
p		1.41	2.11	3.11		as-deposited	
		1.91	1.66	1.08		500 °C, 30 min	
p	Ni	1.31	1.56	1.29	Si-		[8]
	Au	1.35	1.49	1.08			
	Ti	1.94	2.07	1.07			
n	Ti	0.91	–	1.17	Si-	as-deposited	[7]
		~0.94	–	~1.22			
		0.99	–	1.03		500 °C annealed	
		~1.04	–	~1.04			
n	Ni	1.59	–	1.05	Si-		[8]
	Pt	1.39	–	1.01			

thermal expansion coefficients. One example is Ni, which is difficult to evaporate onto SiC, but a thin layer of Ti will act as an adhesion layer. Of course this will result in a Ti Schottky, but it will make the Ni stick, and Ti is actually one of the best metals for all purposes. Sputtered TiW is also a good general purpose metal, for both Schottky and ohmic contacts on n- and p-type SiC.

A high ideality factor is normal for unannealed contacts, but after a 500 °C anneal the ideality factor should be below 1.1. If this is not the case, poor cleaning of the surface or impurity of the metal may be suspected. If the metallisation contains Ti (pure Ti, TiW, TiN, TiC, $TiSi_x$) any remaining silicon dioxide in the interface is reduced to TiO_2 during this anneal, and low ideality factor Schottky contacts should result.

High leakage current for low forward biases can result if ion implantation or reactive ion etching has damaged the SiC surface. In this case, a sacrificial oxidation can improve the contacts. Silicon dioxide is grown thermally prior to metal deposition, but removed by wet etching in the contact region. The thermal oxide surrounding the contact can be left, since this passivates the surface (makes it less sensitive to moisture, etc.). Sacrificial oxidation can also reduce surface roughness, which is another cause of higher leakage currents.

For higher bias, the current from avalanche breakdown occurring from impact ionisation eventually dominates the leakage current. Junction termination structures need to be used to control this type of leakage, and this is further discussed in Chapter 7.

6.3 OHMIC CONTACTS

6.3.1 Introduction

This section concerns ohmic contacts and will start with the theory for on-resistance of metal-semiconductor contacts, and describe different current transport mechanisms, such as thermionic emission (same as Schottky), thermionic field emission and field emission. The formation of ohmic contacts is similar to the formation of Schottky contacts, so only the differences will be described. To form an ohmic contact, as high a doping as possible is required, $10^{19}\,cm^{-3}$ and above.

6.3.2 Tutorial

6.3.2.1 Formation of the ohmic contact

The formation of the ohmic contact follows the same four steps as for Schottky contacts. Cleaning is just as important, deposition and patterning are the same, but annealing is usually performed at higher temperatures, 900 °C or more. Unlike Schottky-contact annealing, a reaction between the metal and the SiC is often sought for a good ohmic contact. This reaction can be promoted if there is some surface damage, so sometimes the surface is damaged deliberately in a controlled fashion. The metal choice is very important, and the reaction products must be investigated. Some metals will form both silicides and carbides, in which case the consumed SiC can be incorporated in the metal alloy. Some metals do not form carbides, in which case carbon

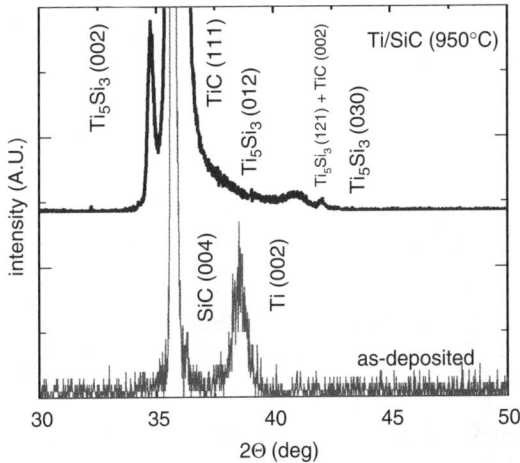

FIGURE 6.6   X-ray spectra of Ti deposited on 4H SiC before and after 950 °C anneal. The deposited Ti is identified by its polycrystalline (002) peak. After reaction, both the Ti_5Si_3 phase and the TiC phase are clearly seen.

precipitates will be present in the metal, and this is detrimental to the specific contact resistance. Some metals do not form either carbides or silicides, and these are difficult to use for ohmic contacts unless very high doping can be used. In any case, the reaction between the metal and the SiC causes roughening of the metal film, which makes wire bonding difficult and can cause reliability problems. The common solution is to deposit only a thin metal film for the reaction, and then deposit additional layers of metal after the high-temperature annealing is finished.

6.3.2.2   Materials characterisation of ohmic contacts

Since chemical reactions take place between the metal and SiC, a chemical analysis of the resulting alloy is of interest. Apart from XRD (FIGURE 6.6) and RBS (FIGURE 6.7), which were discussed for the Schottky contacts in Section 6.2.2 above, several other methods are commonly used. Scanning electron microscopy (SEM) can be used to look at the surface of the metal, and cross sections through the metal/SiC interface can be investigated with SEM (FIGURE 6.8) or transmission electron microscopy (TEM) (FIGURE 6.9). The latter method can achieve atomic resolution, but the sample preparation is time consuming, difficult and expensive since the cross section needs to be thinned to less than 100 nm. Scanning X-ray photoelectron spectroscopy (XPS) can also be used to identify metal phases

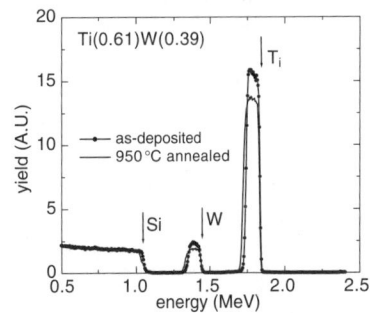

FIGURE 6.7   RBS spectra for as-deposited and 950 °C annealed TiW contacts. The broadening of the peaks and reduction in height indicate reaction with the SiC, as the metal thickness is directly proportional to the width of the peaks. If the metals do not have well-separated masses, this technique is not as useful, and small atomic masses will overlap with the silicon substrate signal.

121

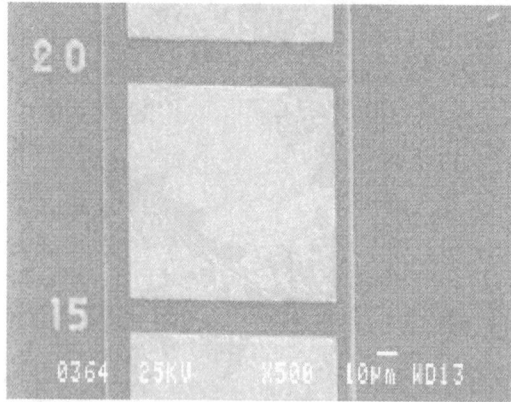

FIGURE 6.8   SEM micrograph of Au/Ti/TiW/SiC contacts after long-term reliability tests for 308 h.

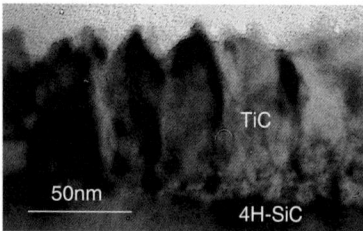

FIGURE 6.9   A TEM cross section of TiC epitaxially deposited on SiC through co-evaporation. A columnar structure can be seen, but the interface is sharp.

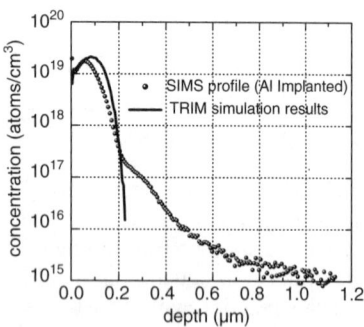

FIGURE 6.10   SIMS depth profile of ion-implanted aluminium, compared to TRIM simulation. A sacrificial layer of silicon nitride was used to achieve a high surface concentration of aluminium.

in the contact. Secondary ion mass spectrometry (SIMS) is also used, mainly to investigate the doping concentration under the contact (FIGURE 6.10). Atomic force microscopy (AFM) can be used to measure the roughness of the metal surface after annealing.

6.3.2.3   Electrical characterisation of ohmic contacts

Earlier, we discussed only one current transport across the Schottky barrier, and that is thermionic emission (TE), described by EQN (1), where electrons are thermally excited over the barrier. When the doping concentration is higher in the semiconductor, other current transport mechanisms can dominate, because the depletion region becomes narrower. If we consider the n-type case, the transport mechanisms can be seen in FIGURE 6.11. In the intermediate doping case, thermionic field emission (TFE) dominates. For high doping, the depletion region is sufficiently thin for direct tunnelling to occur, and this is known as field emission (FE).

The key electrical parameter for ohmic contacts is the specific contact resistance, also known as the contact resistivity. This is measured in Ω cm^2, and the voltage drop across the contact is thus calculated by multiplying by the current density through the contact in A/cm^2. Although theoretical calculations can be made of the specific contact resistance, the equations are quite complicated and difficult to use because of the exponential dependence on Schottky barrier height and doping [9]. FIGURE 6.12 shows the calculated specific contact resistance for high doping, which is the regime where both field emission and thermionic field emission can dominate. It is clearly seen that the doping is very critical, since low Schottky barrier heights cannot normally be expected on SiC. Normally, the specific contact resistance has to be measured for the material and metal used after the annealing process. The annealing process needs to be optimised, since the specific contact resistance is reduced for increased temperature and annealing time, but only up to a certain limit. Special test structures are used for this (see next section).

Long-term stability of the contacts is also an important issue. This is investigated with contact resistance measurements after intervals of annealing in vacuum or oxidizing ambient. A Pt or Au cap layer has been found to increase the stability of TiW contacts (FIGURE 6.13).

6.3.2.4 *Specific contact resistance test structures*

To measure the specific contact resistance, special test structures are needed, because the voltage drop across the contact is very small. One set of probes is normally used to measure the voltage, while the current is forced through another set of probes, to avoid the influence from the voltage drop in the cables and in the probe. Since it is the current that is measured rather than the current density, more advanced test structures use sophisticated means to constrict the current flow to a well-defined cross-sectional area. This will add extra process steps to fabricate the test structure. In all cases, the highly doped layer to which the contacts are made needs to be junction isolated from the substrate (n + epi on p substrate, or p + on n), and the thickness and doping of the layer must be known.

The simplest test structure uses the Kuphal method [10] and consists of circular contacts with diameter d and a fixed spacing s on an epitaxial layer (FIGURE 6.14). Only one mask is needed, to define the contacts. If a current is forced from contact a to d (I_{ad}), and the voltage is measured between a and b (V_{ab}), and b and c (V_{bc}) respectively, the specific contact resistance can be calculated as

$$\rho_C = \frac{A}{I_{ad}}\left(V_{ab} - I_{ad}R_{sp} - V_{bc}\frac{\ln\left(3s/d - 1/2\right)}{2\ln 2} \right) \quad (7)$$

where the spreading resistance R_{sp} can be neglected if the ratio between contact area and epilayer thickness is small, which is often the case. With this simple structure, contact resistivities down to $10^{-4}\,\Omega\,cm^2$ can be measured, but the values are overestimated.

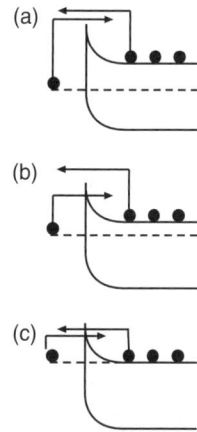

FIGURE 6.11   Energy band diagram for a Schottky contact to a (a) low-doped (TE), (b) intermediately doped (TFE) and (c) highly doped (FE) n-type semiconductor. The arrows indicate the electron flow.

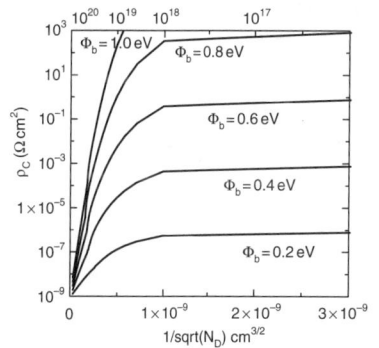

FIGURE 6.12   Calculated specific contact resistance versus doping concentration for barrier heights from 0.2 to 1.2 eV. The calculation uses thermionic field emission (TFE) for doping concentrations between 10^{16} and $10^{18}\,cm^{-3}$, and field emission (FE) for doping between 10^{18} and $10^{21}\,cm^{-3}$.

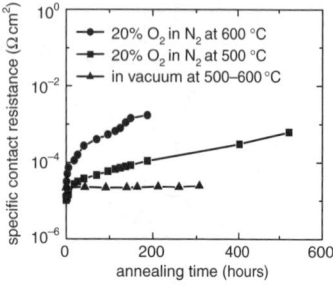

FIGURE 6.13 Specific contact resistance for Pt/Ti/TiW contacts measured after elevated temperature annealing in different ambients.

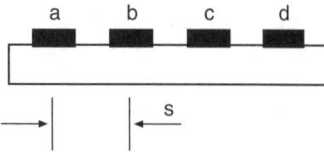

FIGURE 6.14 Schematic view of Kuphal test structure.

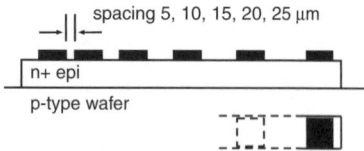

FIGURE 6.15 Schematic side and top view of linear TLM structure.

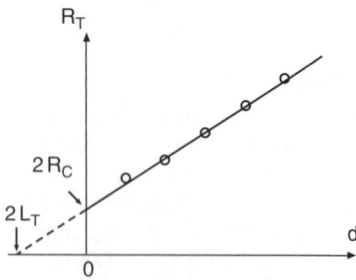

FIGURE 6.16 Plot of total resistance versus spacing.

The most common method is probably the linear transfer length method (L-TLM), sometimes referred to as the transmission line method [11]. An extra process step is added to etch a mesa to constrict the current in one dimension, but increased accuracy is possible so that contact resistivities down to 10^{-6} Ω cm^2 can be measured. The mesa with width Z is contacted through a series of contacts with varying spacing d (FIGURE 6.15). It is important that these contacts reach all the way to the mesa edge, otherwise the measurements will underestimate the specific contact resistance since more current can be conducted at the edges. The total resistance is measured with four probes (two for current, two for voltage) for each of the pairs of contacts, and the total resistance is plotted versus contact spacing. Since the total resistance consists of the contact resistances and the resistance of the epilayer,

$$R_T = \frac{\rho_S d}{Z} + 2R_C \quad \text{and} \quad \rho_C = R_C L_T Z \qquad (8)$$

Here, ρ_S is the sheet resistance of the epilayer, which can be extracted from the plot of R_T versus spacing d: see FIGURE 6.16. L_T is the linear transfer length, which gave the method its name, and it is extracted from the x-axis intercept in the graph. Together with R_C, which is extracted from the y-axis intercept, the specific contact resistance ρ_C is easily determined from EQN (8). The slope in the figure is proportional to the sheet resistance ρ_S of the epilayer.

If extensive reaction occurs between metal and SiC, so that the epilayer is significantly thinner under the contact than between the contacts, the equations need to be modified [12]. If more contacts and spacings can be used on the mesa, the final extracted value is better and less sensitive to measurement errors.

A variation of this method is to use concentric circular contacts and to measure the resistance between the circular contacts [13]. This avoids the mesa-etching step, but due to the large size of the resulting structure, usually only three circles with two spacings can be made. In this case the extraction comes from extrapolating from two resistance measurements in a diagram, which makes this method highly sensitive to measurement errors.

To be able to measure even smaller contact resistivities, the area has to be constricted so that a large enough voltage drop occurs. The Kelvin structure or Kelvin cross-bridge exists in many variations [14]. The basic principle is that the doped layer is isolated by reactive ion etching of a mesa, which is contacted with two large contacts and one small contact (FIGURE 6.17). The contact area is defined by using an insulating layer with a contact hole etched in it. Finally, adding a metal layer for contacts results in three process steps, and the area of the centre contact hole needs to be well defined. The advantage is that this is a direct measurement of specific contact resistance. Current is passed from contact 1 to contact 2, which goes through the contact hole. The voltage is measured across this contact via contact pads 3 and 4. The contact resistance R_C and specific contact resistance ρ_C are then calculated from:

$$R_C = \frac{V_{34}}{I} \quad \text{and} \quad \rho_C = R_C A \qquad (9)$$

where A is the contact hole area. The added process complexity allows us to measure contact resistivities down to $10^{-8}\ \Omega\ cm^2$, if several different areas are made on the same test chip.

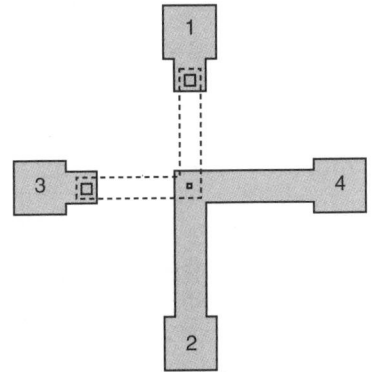

FIGURE 6.17   Schematic view of Kelvin structure; dashed line indicates the etched mesa in the semiconductor.

6.3.3   Comparison of techniques

TABLE 6.2   Review of specific contact resistance measurements on 4H SiC in the literature.

Polytype (n or p)	Metals	Doping (cm^{-3})	ρ_c $(\Omega\ cm^2)$	Face	Annealing	Ref.
4H(n)	Ni-Cr	4.8×10^{17}	1.0×10^{-4} $\sim 1.6 \times 10^{-5}$	Si-	1100 °C, 3 min	[15]
		1.3×10^{19}	1.2×10^{-5}			
4H(p)	Si/Pt Al/Ti	1×10^{19}	$\sim 10^{-3}$ $\sim 10^{-4}$	Si-	30 °C–400 °C 1100 °C, 3 min	[16]
4H(n)	Ni Cr W	10^{17}–10^{18}	10^{-4}–10^{-6}	Si-	1000 °C~ 1050 °C, 5 min	[17]
4H(n) 4H(p)	TiC Ti	1.3×10^{19} $>10^{20}$ $>10^{20}$	4×10^{-5} 6×10^{-5} 8×10^{-4}	Si	950 °C	[18]
6H(n)	Ti	$>10^{20}$	2×10^{-5}	Si	950 °C	[19]
4H(p)	TiC	2×10^{19}	1×10^{-4}	Si	850 °C	[20]
4H(n)[a] 4H(p)[b]	Ni	1×10^{19} 1×10^{21}	6.0×10^{-6} 1.5×10^{-4}	Si	1050 °C 10 min	[21]
4H(n)		1×10^{19}	1.5×10^{-5}		1000 °C, 5 min	[22]
		1.1×10^{19}	7.5×10^{-6}		950 °C, 30 min	[23]
6H(n)		5×10^{19}	1×10^{-6}		1000 °C, 5 min	[24]
		7.8×10^{18}	4–9×10^{-6}		950 °C, 2 min	[25]

TABLE 6.2   (continued)

Polytype (n or p)	Metals	Doping (cm^{-3})	ρ_c (Ω cm^2)	Face	Annealing	Ref.
6H(n)		3.2×10^{17}	3×10^{-6}		1200 °C, 1 min	[25]
		$2–5 \times 10^{18}$	$8–9 \times 10^{-5}$	C	1000 °C	[26]
4H(n)	TiW (30:70)^c	1.3×10^{19}	$2–6 \times 10^{-5}$	Si	950 °C, 30 min	[23,27,28]
4H(p)		6×10^{18} ~$>10^{20}$	1.2×10^{-4} ~4×10^{-6}			
6H(n)	TiW (10:90)^c	7×10^{18}	1×10^{-4}		750 °C, 5 min	[29]
3C(n)		1.7×10^{20}	7.8×10^{-5}		900 °C, 15 min	[30]
6H(n)		6×10^{18}	3.4×10^{-4}			

[a] Nitrogen was implanted.
[b] Aluminium and carbon were co-implanted.
[c] Weight ratio (Ti:W).

TABLE 6.3   Comparison of different test structures for specific contact resistance.

Method	Type	Number of masks	Measurement limit
Kuphal	indirect	1 (contact)	10^{-4} Ω cm^2
TLM	indirect	2 (mesa, contact)	10^{-6} Ω cm^2
Kelvin	direct	3 (mesa, isolation, contact)	10^{-8} Ω cm^2

6.3.4   Overcoming problems

Many problems occur when making ohmic contacts to SiC, especially to p-type since the dopant ionisation is low. A very high dopant concentration is needed to reliably form ohmic contacts. With highly doped epitaxial layers this can usually be done if the doping is higher than 5×10^{18} cm^{-3} for n-type, and 10^{19} cm^{-3} for p-type. Ion-implanted layers are more complicated since it is difficult to achieve a high surface concentration, but sacrificial layers can be added during the ion implantation and then removed [20].

The standard solution for a specific contact resistance that is too high is increased annealing, either for longer time or at higher temperatures. Cleaning before metal deposition is, of course, important, but perhaps more important is the selection of a metal that reacts cleanly with SiC. Ni and Co are examples of metals that do not form carbides, and hence carbon clusters can be expected in the contacts, which increases the specific contact resistance. If metal layers that are too thick are deposited for the alloying, the metallisation will be rough after the reaction of metal with SiC. It is better to deposit a thin layer first, and add a thicker layer after the annealing. Stoichiometric contacts can also be deposited to form $CoSi_2$ and other silicides [31]. Reducing the annealing temperature also lessens the roughening.

For long-term stability of the contacts, capping is needed. To finally contact a device, a bondable metal needs to be deposited on top. It has been found that Au or Pt, which are good metals for wire bonding, also protect the underlying Ti-based contacts, and hence long-term reliable contacts can probably be achieved using this structure [23].

6.4 RECENT DEVELOPMENTS AND FUTURE TRENDS

Au nano-particles (radius 10 nm) deposited prior to Ti have been found to lower the Schottky barrier height [19]. This is believed to be the result of a combination of field crowding at the nano-particles, and the large difference in metal work function between Ti and Au. The lowering is about 0.2 eV on both n-type and p-type, and both 4H and 6H. However, this lowering is too small to significantly affect the specific contact resistance of ohmic contacts.

Epitaxial metal films of TiC co-evaporated from Ti and C_{60} have proved to be an interesting way to achieve ohmic contacts as-deposited at 500 °C [18]. Good Schottky contacts can also be made with this alloy.

For most devices, ohmic contacts are needed to both n-type and p-type SiC. TiW has been found to work well without excessive annealing (980 °C), and this will simplify processing tremendously [27]. Although Ni has a slightly lower specific contact resistance on n-type, Ni contacts are rough after annealing at high temperatures and the reliability is questionable.

6.5 CONCLUSIONS

The properties of metal contacts to SiC are summarized in TABLE 6.4.

Schottky contacts to SiC are easy to fabricate, since we naturally will achieve a high barrier due to the wide bandgap, if the

TABLE 6.4 Summary of contact formation to SiC.

Schottky	Ohmic
High barrier	Low barrier
Low doping $<10^{17}$ cm^{-3}	High doping $>10^{19}$ cm^{-3}

doping is not too high. If the annealing temperature is not too high, there is no problem, since there should be no reaction. This means that all ohmic contacts have to be annealed before the Schottky contact metal is deposited.

As long as the doping is high enough, especially for p-type, and high annealing temperatures are allowable in the process, good ohmic contacts can be made. High enough in this case means more than $10^{19}\,cm^{-3}$ approximately, and temperatures of around 900°C. Ion implantation can be used to increase surface concentration, but care has to be taken with the annealing process so that the doped layer is not accidentally removed through etching, oxidation or reaction with the metal.

REFERENCES

[1] E.H. Rhoderick, R.H. Williams [*Metal-Semiconductor Contacts* 2nd edn (Clarendon Press, 1988)]
[2] S.M. Sze [*Physics of Semiconductor Devices* 2nd edn (John Wiley & Sons Inc., 1981)]
[3] J. Bardeen [*Phys. Rev. (USA)* vol.71 (1947) p.717]
[4] J. Crofton, S. Sriram [*IEEE Trans. Electron Devices (USA)* vol.43 (1996) p.2305]
[5] K.J. Schoen, J.M. Woodall, J.A. Cooper Jr., M.R. Melloch [*IEEE Trans. Electron Devices (USA)* vol.45 (1998) p.1595]
[6] A. Itoh, H. Matsunami [*Phys. Status Solidi A (Germany)* vol.162 (1997) p.389]
[7] D. Alok, R. Egloff, E. Arnold [*Mater. Sci. Forum (Netherlands)* vol.264–268 (1998) p.929–32]
[8] V. Saxena, J.N. Su, A.J. Steckl [*IEEE Trans. Electron Devices (USA)* vol.46 (1999) p.456]
[9] T.P. Chow, R. Tyagi [*IEEE Trans. Electron Devices (USA)* vol.41 (1994) p.1481]
[10] E. Kuphal [*Solid-State Electron. (UK)* vol.24 (1981) p.69]
[11] H.H. Berger [*Solid-State Electron. (UK)* vol.15 (1972) p.145]
[12] G.K. Reeves, H.B. Harrison [*IEEE Electron Device Lett. (USA)* vol.EDL-3 (1982) p.111]
[13] G.K. Reeves [*Solid-State Electron. (UK)* vol.23 (1980) p.487]
[14] D.K. Schroder [*Semiconductor Material and Device Characterization* 2nd edn (John Wiley & Sons Inc., 1998)]
[15] E.D. Luckowski et al [*J. Electron. Mater. (USA)* vol.27 (1998) p.330]
[16] N.A. Papanicolaou, A. Edwards, M.V. Rao, W.T. Anderson [*Appl. Phys. Lett. (USA)* vol.73 (1998) p.2009]
[17] T. Lamp, S. Liu, M.L. Ramalingam [*AIP Conf. Ser. (USA)* vol.387 (1997) p.359–64]
[18] S.K. Lee, C.M. Zetterling, M. Östling, J.P. Palmquist, H. Högberg, U. Jansson [*Solid-State Electron. (UK)* vol.44 (2000) p.1179]
[19] S.-K. Lee et al [*Mater. Sci. Forum (Switzerland)* vol.389–393 (2002) p.937–40]
[20] S.K. Lee et al [*Appl. Phys. Lett. (USA)* vol.77 (2000) p.1478]

[21] L.G. Fursin, J.H. Zhao, M. Weiner [*Electron. Lett. (UK)* vol.37 (2001) p.1092]

[22] C. Hallin et al [*J. Electron. Mater. (USA)* vol.26 (1997) p.119]

[23] S.-K. Lee, C.-M. Zetterling, M. Ostling [*J. Appl. Phys. (USA)* vol.18 (2002) p.201–9]

[24] T. Uemoto [*Jpn. J. Appl. Phys. (Japan)* vol.34 (1995) p.L7–9]

[25] J. Crofton, P.G. McMullin, J.R. Williams, M.J. Bozack [*J. Appl. Phys. (USA)* vol.77 (1995) p.1317]

[26] M.G. Rastegaeva et al [*Proc. 6th SiC Related Materials Conference* Kyoto, Japan (IOP Publishing Ltd., 1995) p.581]

[27] S.K. Lee, C.M. Zetterling, M. Östling, J.P. Palmquist, U. Jansson [*Microelectron. Eng. (Netherlands)* vol.60 (2002) p.261–8]

[28] S.K. Lee, S.M. Koo, C.M. Zetterling, M. Östling [*J. Electron. Mater. (USA)* vol.31 (2002) p.340–5]

[29] J. Crofton, J.R. Williams, M.J. Bozack, P.A. Barnes [*Inst. Phys. Conf. Ser. (UK)* no.137 (1993) p.719]

[30] J. Kriz, K. Gottfried, T. Scholz, C. Kaufmann, T. Gebner [*Mater. Sci. Eng. B (Netherlands)* vol.46 (1997) p.180–5]

[31] N. Lundberg, M. Östling [*Solid-State Electron. (UK)* vol.38 (1995) p.2023]

Chapter 7

Devices in SiC

C.-M. Zetterling, S.-M. Koo and M. Östling

7.1 CHAPTER SCOPE

This chapter is about the devices that can be made in SiC. The first section discusses process integration, which is important for all device types. The chapter is thereafter divided into sections after device application, since the different applications have similar designs and problems even though the device types used are quite different. The sections cover high-voltage devices, high-frequency devices, and finally high-temperature and optical devices. The purpose of this chapter is to show which device types are most popular in SiC, and to compare their typical cross sections with each other. For details on the device operation and modelling equations, the reader is referred to other books dedicated to this subject [1,2].

7.2 PROCESS INTEGRATION

7.2.1 Introduction

This section will give a brief tutorial on process integration, which needs to be kept in mind for all device designs that can be made. A lateral MOSFET process in SiC will be used as an example. Note that the device cross sections are not drawn to scale; the vertical dimension is exaggerated.

7.2.2 Tutorial

Interdependencies between process steps proscribe a strict order of process steps to be used. Some materials deposited in later steps cannot withstand the high temperatures used early in the process. The damage created in some of the process steps must be removed before further processing, otherwise unpredictable results will occur. Contact annealing and oxidation are steps in which the reaction rate is very sensitive to SiC surface damage. Starting from the choice of wafer, and finishing with the metallisation, let us proceed step-by-step.

7.2.2.1 Choice of bulk wafer

The price of wafers can vary substantially, and generally n-type is cheaper than p-type, which in turn is cheaper than semi-insulating wafers. For lateral devices it can be worth investigating the use of an n-type wafer with a thick p-type buffer layer, instead of a p-type or semi-insulating wafer (FIGURE 7.1). For vertical devices it is important that the wafer has high doping, so that the series resistance is low and backside contacts can be made without problem. Wafers with fewer defects and larger usable area are, of course, more expensive, but for experimental runs and small-area devices the highest-quality wafers are not needed. Wafers are normally purchased from one of the commercial vendors, since developing bulk SiC production is a major challenge in itself; see Chapter 2.

7.2.2.2 Epitaxy for low-doped regions

Normally, any device will include a low-doped blocking layer, also called the drift layer. The higher the blocking voltage, the thicker this drift layer needs to be and the lower the doping; see calculations in Chapter 1. At least one epitaxial layer is needed for all devices, since the substrate wafer has too high a doping to support any voltage. For lateral devices, a low-doped channel layer with well-controlled doping and thickness is needed (FIGURE 7.2). For thick epi, high-temperature chemical vapour deposition (HTCVD) can be an alternative to standard CVD; see Chapter 2. On semi-insulating wafers a buffer layer is needed below the channel epilayer to avoid traps from the semi-insulating interface influencing the threshold voltage. Sometimes a buffer layer of the same type and doping concentration is grown on the wafer. If possible, all epitaxy should be grown in one run without breaking the vacuum. However, since high doping concentrations will contaminate the reactor, the epitaxy steps are usually divided into low-doped epitaxy, and high doping of different dopants.

7.2.2.3 Epitaxy (and etching) and/or ion implantation for high-doped regions

Once a low-doped region has been grown, the remaining regions can, in principle, be defined by ion implantation. However, some regions are not suitable for ion implantation, since there can be

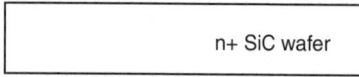

FIGURE 7.1   Starting material is an n-type SiC wafer.

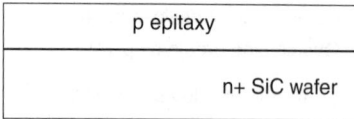

FIGURE 7.2   The p-channel epitaxy is grown.

remaining damage even after the annealing, which influences the lifetime. One example is the base region of the bipolar transistor, which needs a long lifetime, and therefore it is not recommended to ion implant either the intrinsic base or the emitter. For ohmic contact regions, epitaxy can often create higher doping at the surface than ion implantation can. If this is done after some etching or ion implantation, this is called regrowth, and it has to be done on a bare SiC surface, without metals or dielectrics. Ion implantation can be done selectively over the surface using a mask (FIGURE 7.3), but epitaxial growth will cover the whole surface. Therefore etching with a mask is needed to define regions when epitaxy is used.

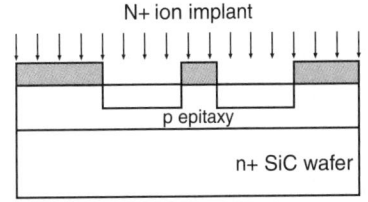

FIGURE 7.3   The source and drain n+ regions are ion implanted with an oxide mask.

7.2.2.4   *Ion-implantation masks and annealing*

To reduce the damage from ion implantation, the wafer is normally heated to 500–700 °C during the ion implantation. This means that photoresist masks cannot be used; instead silicon dioxide or metal masks are used. After ion implantation, a yet higher temperature is needed for the annealing of crystal damage, around 1200 °C for n-type and up to 1700 °C for p-type (FIGURE 7.4). This, of course, means that no metal or dielectrics can be present on the surface, hence all ion implantation is done early in the process. To make the removal of metal masks used for ion implantation easier, a dielectric like silicon dioxide or silicon nitride can be deposited first, and the dielectric and metal can then be removed by lift-off using HF. More details on ion implantation can be found in Chapter 3.

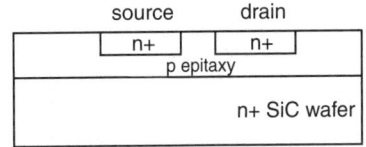

FIGURE 7.4   The ion implant anneal at 1200 °C activates the source and drain.

7.2.2.5   *Etching*

Reactive ion etching using an inductively coupled plasma (ICP) is highly recommended: see Chapter 4. ICP allows high etch rates, and can be tuned to yield low damage. If the ICP etcher is run at low platen power, damage is reduced, and mask erosion can be kept small. Mask choice is very important. Metal masks like Al or Ni have the highest selectivity to etching (more than a factor of ten), especially Al which can be oxidised in situ by switching an oxygen plasma with the etching plasma. The metal masks can be deposited on an oxide or nitride layer (deposited by PECVD) to reduce stress and make removal easier by HF (FIGURE 7.5). Dielectric masks like silicon dioxide have less selectivity; normally the thickness needs to be about twice the thickness of the SiC to be etched. A photoresist mask can only be used for very small etch depths, but is the easiest for processing, since it needs no extra patterning. Since both oxide masks and photoresist masks are etched during the process, the dimensions

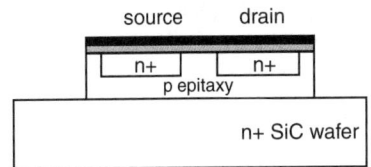

FIGURE 7.5   The isolation mesa is dry etched.

change and straight edges cannot be preserved. On the other hand, there is a risk of trenching or mask residues seen when using metal masks.

7.2.2.6 Sacrificial oxidation

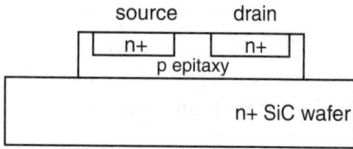

FIGURE 7.6 Sacrificial oxidation is performed to remove etching damage.

Sacrificial oxidation can be used to remove SiC through thermal oxidation, and the silicon dioxide is then removed by wet etching with HF (FIGURE 7.6). This thermal oxide may also be left on the parts of the SiC without metal to passivate the surface. However, in some labs the SiC wafers are not allowed into the oxidation furnace if there has been a metal mask present on their surfaces at any stage before (even if deposited on a sacrificial dielectric) because of the risk of contamination. Either only oxide masks for ion implantation may be used, and oxide or photoresist for etching, or a separate oxidation furnace has to be used.

7.2.2.7 Gate dielectric for MOSFETs and IGBTs

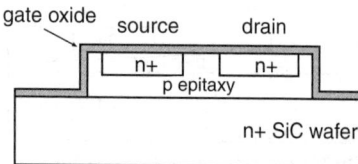

FIGURE 7.7 The gate oxide is grown after removing the sacrificial oxide.

Thermally grown silicon dioxide seems to be the main choice for gate dielectrics, but also stacks of oxide/nitride/oxide may be deposited and annealed (FIGURE 7.7). Apart from the growth, the annealing process is critical for a good gate dielectric. The details and associated difficulties are discussed at length in Chapter 5. One thing to remember is that as with sacrificial oxidation, there can be a contamination issue using oxidation furnaces.

7.2.2.8 Gate material

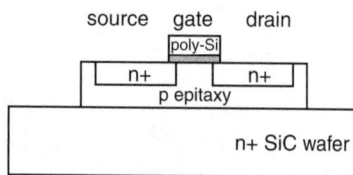

FIGURE 7.8 Polysilicon is deposited and patterned for the gate.

For silicon VLSI MOSFETs, polysilicon has been the sole material choice ever since the self-aligned process was invented. To align the source and drain with the gate, the polysilicon gate is deposited and patterned before source and drain implantation. However, this means that the gate is present during the high-temperature anneal after ion implantation. It is doubtful if even the n-type anneal at 1200 °C would allow a polysilicon gate to remain unaffected. Therefore, the gate is deposited after all ion implants have been made in the SiC devices, without self-alignment (FIGURE 7.8). The choice between polysilicon and metal gate is therefore not clear, and both choices are possible, depending on the device requirements and facilities present. The advantage with polysilicon is that it can withstand the high-temperature ohmic contact anneal better than most metals.

7.2.2.9 Ohmic and Schottky contacts

A thin film of metal is deposited and patterned in some way, according to procedures described in Chapter 6. If there is an oxide on the SiC surface it has to be removed before the metal is deposited (FIGURE 7.9). After a high-temperature anneal (around 900 °C) the metal is alloyed to the SiC. Backside metallisation is also performed at this stage. Thereafter, the Schottky metal is deposited and patterned, and a lower-temperature anneal is used (500 °C).

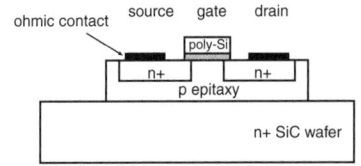

FIGURE 7.9 Ohmic contacts are deposited, patterned and annealed.

7.2.2.10 Intermetal isolation

Silicon dioxide or silicon nitride is deposited over the metal. Contact holes can then be etched with some process with high selectivity to the contact metal or polysilicon gate. Wet etching can be used for silicon dioxide using buffered HF, if the contact holes are not too small (FIGURE 7.10). For air bridges, the dielectric is left out and instead sacrificial photoresist is deposited and patterned.

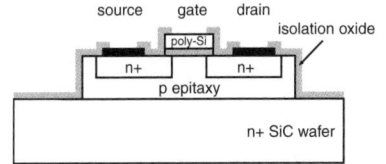

FIGURE 7.10 Isolation oxide is deposited and contact holes are etched.

7.2.2.11 Thick metal

Since only a thin layer of metal is deposited for the ohmic contacts, to avoid roughening and carbon precipitation during the contact anneal (see Chapter 6), a thick metal layer needs to be deposited for probing or bonding (FIGURE 7.11). Metal layers may also be needed to reduce the resistivity of gates, and to interconnect several devices on the same chip. If the underlying metal surface is clean, annealing is normally not needed. The top layer for wire-bonding may need some extra consideration, since some equipment can only bond to Au or Pt that has not been heat treated. For air bridges, thick Au is electroplated on a sputtered seed, and when finished the supporting photoresist can be removed with acetone. The free-standing metal interconnects are used for high-frequency devices where the capacitance has to be minimised (see Section 7.4.2.2).

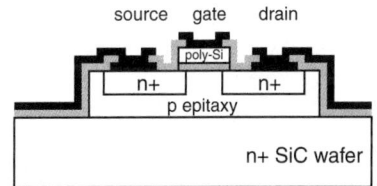

FIGURE 7.11 Interconnects are deposited and patterned.

7.2.2.12 Via holes

For high-frequency devices there is a need for contacting the source to the backside, even though the device is lateral. This increases the gain by reducing the source inductance, which otherwise causes negative feedback of the gate voltage. This is done by via holes etched through the entire wafer, after first thinning the wafer by grinding (FIGURE 7.12). The process steps added are (a) wafer thinning, (b) backside mask deposition and lithography, (c) ICP deep etching (50–100 μm) and (d) electroplating of the via holes. This is normally done before forming the air bridges.

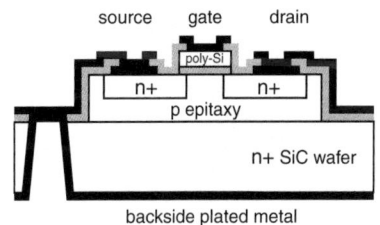

FIGURE 7.12 For high-frequency devices via holes are etched and plated from the back side.

7.2.3 Overcoming problems

The high temperatures that are needed for processing of SiC during growth, ion implantation, annealing and oxidation are probably the biggest difficulty. Most of the equipment will have to be tailor-made for SiC. The issue of contamination may also force the purchase of dedicated equipment.

Another problem, which is not apparent until processing starts, is that SiC wafers are much smaller than standard wafers of Si or even GaAs. The high price per wafer also means that many experiments are performed on smaller parts of one wafer. Wafer-handling equipment of most commercial machines is made for four-inch wafers or larger, so special holders may be needed to mount the small SiC pieces. The end station of the ion implanter needs to heat the small sample as well. Some furnaces allow smaller pieces to be loaded manually. In some etching machines the SiC pieces can be fixed to a Si wafer of the correct size with photoresist. Metal-deposition systems are sometimes a little more flexible, unless a load-lock is used. Spinning on photoresist on small samples will need a specially made chuck. Lithography is difficult, as newer systems have no manual loading and can only handle standard-sized wafers. If the SiC piece is mounted on a wafer or holder it may not be flat enough for even exposure. Many research groups use older aligning equipment, which can handle smaller substrates manually. Cleaning and wet etching have to be done manually in beakers under a fume hood or with small Teflon baskets, but one has to be careful not to lose the SiC sample down the drain or on the floor.

7.3 HIGH-VOLTAGE DEVICES

7.3.1 Introduction

This section is about high-voltage devices in SiC. After a brief description of device operation, the critical design points are covered. Some different design variations will be shown, and some different devices will be compared to each other in terms of high-voltage performance. Note that the device cross sections are not drawn to scale; the vertical dimension is exaggerated.

7.3.2 Tutorial

The design of high-voltage devices is based on the fact that SiC can handle about a ten-times higher critical electric field, and therefore will have much lower on-resistance compared to a Si device with the same blocking voltage. A wide drift region needs to be included to support the depletion region from the high reverse voltage. If the drift region is slightly wider than the depletion region at breakdown (when the electric field at the p-n junction reaches the critical field) this is called a non punch-through design (NPT). The drift region can also be made slightly shorter and with higher doping, so that the critical field is not reached when the entire drift region is depleted, and this is called a punch-through (PT) design [2]. Almost all of these devices are vertical, meaning that the main current flow is vertical, from the front of the wafer to the back, and that the high electric field is vertical. Normally, a cell design is used, so that a large-area device actually consists of several smaller devices connected in parallel, either on one chip or using several chips. Therefore several figures show half-cells in such a design, without the termination structures. Growth of thick layers with well-controlled doping (small lateral variation) is very important for these devices, otherwise hot spots can occur and cause the current to be shared unequally between paralleled devices.

7.3.2.1   Termination and passivation

All high-voltage devices have to have structures for electric-field termination and surface passivation. The termination structure is made to lessen the effects of field crowding at the edge of the device, since the electric field can become many times higher at the corners than in the centre of the depletion region. This is illustrated in FIGURE 7.13, and the field crowding is worse in the three-dimensional case. To avoid early breakdown at the edge, the field needs to be distributed over the surface, which costs large area, up to $100\,\mu m$ outside the edge. Some popular techniques are:

- Junction termination extension (JTE) where additional implants with lower doses are made outside the contact, and the dose is critical (FIGURE 7.14).
- Ar or other neutral implants, without annealing the damaged area. The damaged area is resistive and the leakage current spreads out the voltage (FIGURE 7.15).
- Field plates that spread the electric field through capacitive coupling (FIGURE 7.16).
- Field rings, either ion implanted (FIGURE 7.17(a)) or metal (FIGURE 7.17(b)), which are not connected to the anode.

Although these techniques are shown for an ion implanted p-n junction, they are just as valid for mesa-etched epitaxial structures. It is also important to lower and control the surface charge

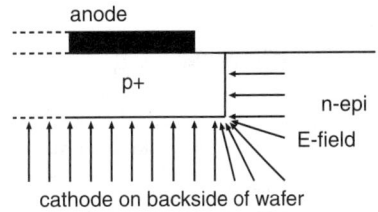

FIGURE 7.13   Field crowding at the p-n junction corner.

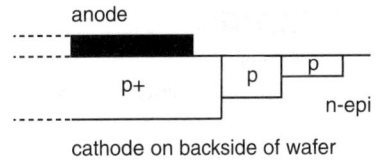

FIGURE 7.14   Junction termination extension.

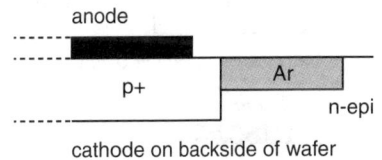

FIGURE 7.15   Ar implanted damaged area.

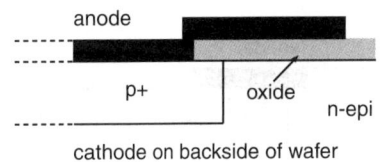

FIGURE 7.16   Field plates on silicon dioxide.

137

(a)

cathode on backside of wafer

(b)

cathode on backside of wafer

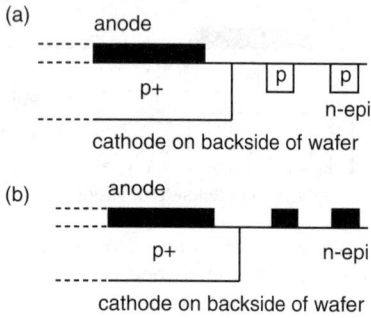

FIGURE 7.17 Field rings surrounding the device.

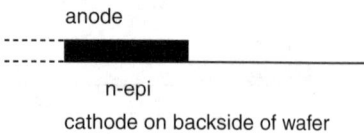

cathode on backside of wafer

FIGURE 7.18 Schottky diode.

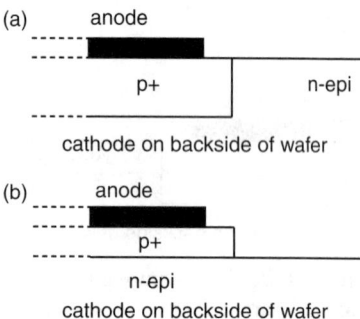

(a)

cathode on backside of wafer

(b)

cathode on backside of wafer

FIGURE 7.19 p-i-n diode.

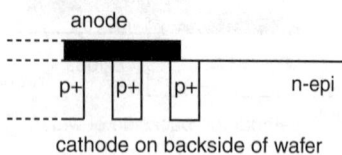

cathode on backside of wafer

FIGURE 7.20 JBS diode.

to avoid field crowding. Surface passivation by, for example, thermal oxidation is used to neutralize surface charges.

7.3.2.2 Diodes

Schottky diodes are the simplest devices; they consist of only a metal contact on the SiC (FIGURE 7.18). A suitable metal on n-type SiC forms a rectifier similar to the p-n junction, and the same thing happens on p-type SiC; see details in Chapter 6. They are very fast since no minority carrier charge is injected, and the on-state voltage (close to the Schottky barrier height) can be low compared to the p-n diode (close to the built-in voltage). For high-voltage operation, a field termination is needed, since the reverse leakage current will be high due to the high electric field.

For higher voltages p-n diodes are used with a very low-doped drift region, almost intrinsic, hence these devices are called p-i-n diodes (i for intrinsic). Although the built-in voltage is higher for SiC (around 2.5 V, since it scales with the bandgap), conductivity modulation by minority carrier injection lowers the on-resistance. These devices are slower than Schottky diodes because the injected charge has to recombine before a voltage can be blocked, but a p-n junction can block higher voltages and has less leakage current. The highly doped region can be made with ion implantation (FIGURE 7.19(a)) or epitaxy (FIGURE 7.19(b)).

A good compromise between the two diode types is the junction barrier Schottky (JBS) diode, which is a Schottky diode with an integrated p-doped grid (FIGURE 7.20). The Schottky regions of the anode conduct forward current with a low voltage drop, and in reverse the p-doped regions pinch off the Schottky area to avoid premature breakdown. The spacing between the p-regions is a critical design parameter; it should be small enough that the leakage current is not excessive before pinch-off occurs. What is gained in forward-voltage drop is lost in extra area needed compared with pin diodes. Compared with Schottky diodes, the large on-resistance due to the pinched-off regions between the p-doped regions is compensated by larger breakdown voltage. The process complexity of the JBS is not much higher, if we take into account that all diodes need junction termination as well.

7.3.2.3 Unipolar transistors

All unipolar transistors are easy to connect in parallel, since they have positive temperature dependence. If one transistor carries more current than the others, this device heats up and the on-resistance increases in the channel, which serves to lower the current in this device. There is no built-in voltage, since the current does not traverse any p-n junction. Since the current consists entirely of majority carrier devices, no conductivity modulation can occur to lower the forward-voltage drop, hence switching is fast. However, for higher blocking voltages the on-resistance of the drift region is too high, and bipolar devices are preferred. The devices normally rely on electron current, since the electron mobility is much higher than the hole mobility in SiC.

The junction field-effect transistor (JFET) is normally made in n-type SiC, with a reverse biased p-type gate. When the reverse bias is increased on the gate, the depletion region of the p-n junction gate will pinch off more of the channel. These devices are normally designed with a wide channel to reduce the on-resistance. The devices are normally-on (negative pinch-off voltage) and conduct current with zero volts on the gate. The junction should not be forward biased, but the high built-in voltage of SiC allows positive voltages of up to about 2 V. The lateral design with dual gates is common (FIGURE 7.21(a)), although a simpler design is to exclude the front gate, and halve the channel thickness, to form a buried gate device [3] (FIGURE 7.21(b)). A vertical design (VJFET) is also possible using regrowth of n-type epilayers on top of p-type ion-implanted regions [4] (FIGURE 7.22(a)). This device suffers from a large gate-to-drain overlap capacitance, just like the lateral JFETs in FIGURE 7.21. A better design (FIGURE 7.22(b)) has the front gate shielded from the drain and therefore has faster switching [4]. In principle, the front side p-n junction gate can be replaced with a Schottky junction (metal) gate, in which case it is called a metal-semiconductor field-effect transistor (MESFET). Due to higher leakage current in the Schottky gate, this design is more common for high-frequency transistors, which operate at lower voltages; see Section 7.4.2.

The metal-oxide-semiconductor field-effect transistor (MOSFET) relies on a metal or doped polysilicon gate isolated with an oxide or other insulator, to control the field in the channel. The depletion mode MOSFET works just like the JFET or MESFET, in that the depletion region under the gate controls the channel width, and this is a normally-on device (source and drain are doped the same type as the channel). More popular is the inversion-mode MOSFET, where the gate voltage is used to bend the bands in the semiconductor so much that inversion occurs. An electron channel is formed in the p-type channel region to connect the source and drain; see FIGURE 7.11. (Note that normally a fourth contact to the p-channel region, called

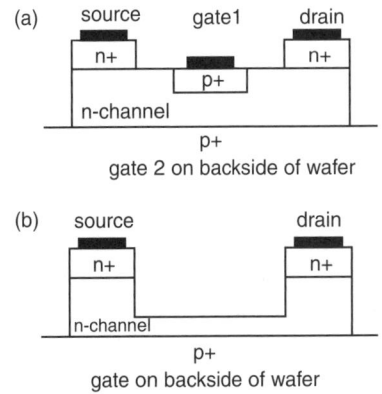

FIGURE 7.21 (a) Lateral JFET with dual gates; (b) lateral JFET with buried gate.

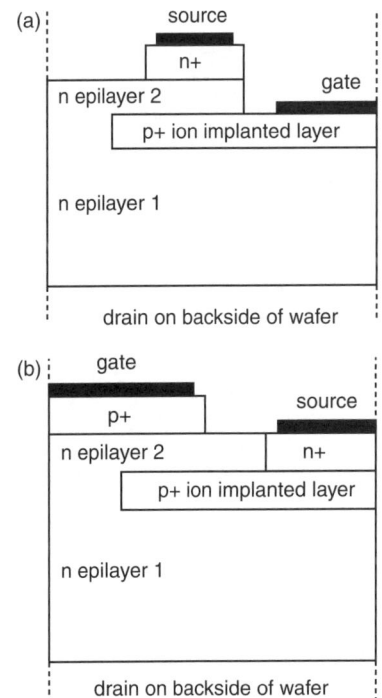

FIGURE 7.22 (a) Vertical JFET (half-cell); (b) vertical JFET (half-cell) with front gate shielded from drain.

body, is needed to ensure that the potential does not drift.) A hybrid between these types is the accumulation-mode MOSFET (or ACCUFET), which has a thin n-type layer on a p-type layer. With zero bias on the gate, the thin n-channel is depleted by the p-type region, but with positive bias the n-channel is accumulated. The on-resistance is high, since the n-channel is rather thin, but the advantage is that the trap situation in the gate oxide is better on n-type than on p-type SiC and the mobility is higher; see Chapter 5. If there is a choice, the inversion channel MOSFET is preferred for high-voltage applications (since it is normally-off), if the inversion mobility can be made high enough.

A common problem for all MOSFETs is the electric field in the oxide. Since the displacement field is continuous across the SiC/insulator interface, the electric field scales with the dielectric constant. The relative dielectric constant of silicon dioxide is 2.5-times lower than for SiC; thus the electric field is higher than in the SiC by a factor of 2.5. When the critical field is reached in the depleted SiC, the field in the oxide is around 6.5 MV/cm ($E_{ox,off}$ in EQN (1)). Although this is not higher than the breakdown field of 10–12 MV/cm, it is too high. For reliability concerns, the oxide field should be kept below 3–4 MV/cm, depending on temperature. This could limit the maximum usable electric field in SiC to below the critical electric field. Another factor to take into account is the electric field in the oxide in the on-state, and it is calculated by dividing the maximum gate voltage used by the oxide thickness ($E_{ox,on}$ in EQN (1)).

$$E_{ox,off} = E_{SiC} \frac{\varepsilon_{r,SiC}}{\varepsilon_{r,ox}}$$
$$E_{ox,on} = \frac{V_{gate}}{t_{ox}}$$

(1)

The lateral MOSFET in FIGURE 7.11 is unable to block very high voltages, since the high drain voltage will deplete and punch through the channel. The LDMOSFET (L for lateral, D for double-diffused) can handle higher voltages using a longer drift region, but at the cost of higher on-resistance [5] (FIGURE 7.23). This design is also popular for high-frequency devices. However, it is easier to manufacture the long low-doped drift region vertically, like the diodes. In Si technology the DMOSFET has been popular, where the D stands for double-diffused, referring to how the gate length is determined by the different diffusion lengths of the n- and p-type regions. Since diffusion cannot occur with masks in SiC (see Chapter 3), the same device is sometimes

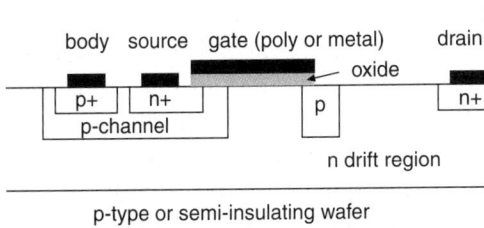

FIGURE 7.23   LDMOSFET (half-cell).

referred to as the DIMOSFET, where DI stands for double-implanted (FIGURE 7.24).

The UMOSFET is also a vertical device, which takes its name from the U-shaped grooves, which are etched to define the gate regions (FIGURE 7.25). Note that the channel is vertical in these devices, and defined by epitaxy rather than lithography, which is the case for the DMOSFET. Another advantage of the UMOSFET is that there is no pinched-off region, which can increase the on-resistance in DMOSFETs. After etching the U-grooves, an epitaxial layer can be regrown to reduce the damage, or to make the device accumulation mode. A problem for UMOSFETs is that the gate oxide thickness will vary because of the different growth rates on different crystallographic faces of SiC; see Chapter 5. One solution is to deposit polysilicon, which is oxidised at a lower temperature so that the process is self-limiting [5]. If a p-type region is ion implanted beneath the groove, the electric field can be reduced in the gate oxide. Otherwise, due to field crowding the electric field can be quite high at the corner and cause breakdown [5].

An interesting hybrid is the JFET/MOSFET cascode connection in FIGURE 7.26. The JFET blocks the voltage, and the MOSFET is used to reverse bias the JFET gate. This places lower demands in terms of breakdown voltage on the MOSFET, and Si MOSFETs can be used. The advantage is that a high-voltage normally-off device is achieved, with MOS gate control [4]. Two devices that integrate this cascode in a novel design are the SIAFET [6] (static induction injected accumulated field-effect transistor) and the SEMOSFET [7] (static channel expansion MOSFET).

7.3.2.4   Bipolar transistors

To lower the on-resistance of the low-doped drift region, bipolar current transport is needed. Through carrier injection, conductivity modulation is achieved. In the bipolar junction transistor

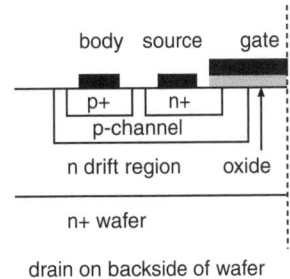

FIGURE 7.24   DIMOSFET (half-cell).

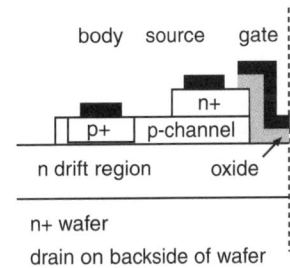

FIGURE 7.25   UMOSFET (half-cell).

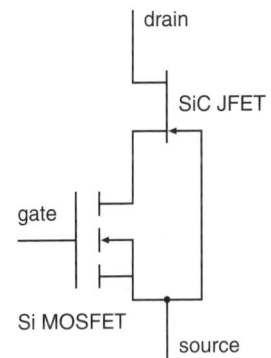

FIGURE 7.26   The MOSFET/JFET cascode.

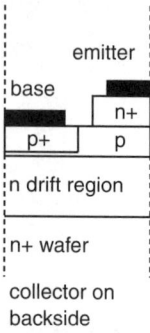

FIGURE 7.27 Bipolar junction transistor (half-cell).

(BJT), two p-n junctions are connected in series. The collector-base junction is reverse biased, and the base-emitter junction is forward biased. In the following, an npn transistor will be considered, FIGURE 7.27. When a hole charge is injected in the base, the barrier is lowered and more electrons are injected into the base from the emitter. If the base width is shorter than the minority carrier diffusion length, most of the electrons reach the collector drift region. The built-in voltages of the two p-n junctions cancel, and a low forward voltage drop can be achieved. An important figure of merit is the current gain β, measured as the ratio between the collector current and the base current. The current gain is proportional to the ratio between doping in the emitter and the base (EQN (2)):

$$\beta = \frac{I_C}{I_B} = \frac{N_E}{N_B} \tag{2}$$

To establish the total on-state loss one has to take the loss in the base into account. Although the base current is lower than the collector current by the factor beta, the built-in voltage for the base-emitter is around 3 V. To further increase the current gain, a heterojunction between emitter and base can be used. This is called a heterojunction bipolar transistor (HBT), and a GaN or AlGaN emitter on the SiC base has been proposed. In this case, the current gain increases by an exponential factor of the valence band offset between the two materials:

$$\beta = \frac{I_C}{I_B} = \frac{N_E}{N_B} e^{\Delta E_g/kT} \tag{3}$$

A wide or highly doped base is needed to support the reverse collector voltage, otherwise the base will be punched through. For the BJT this will reduce the gain, but for the HBT this is not as large a problem.

For Si BJTs and HBTs the Kirk effect, also called base push-out, is a problem. This occurs if the collector current density is higher than the doping in the collector, EQN (4) (v_{sat} is the saturated electron velocity). Since the wide bandgap of SiC allows a higher collector doping for the same breakdown voltage, the threshold for the Kirk effect is almost 100 times higher.

$$J_C \geq q\, N_C\, v_{sat} \tag{4}$$

Another advantage of SiC bipolar transistors over their Si counterparts, is that the temperature coefficient can be positive in some temperature regions. Since the p-type dopant is not fully ionised at room temperature (see Chapter 3), the current gain is larger at room temperature. Divide EQNS (2) and (3) by the ionisation in the base. At higher temperatures, the current gain is reduced both due to increased ionisation of the p-type dopant, and due to the reduced mobility in the base (reduced diffusion length). Therefore, the SiC BJTs may be paralleled just like the MOSFETs. A Darlington connection can also be considered (FIGURE 7.28). Here, the combined current gain is high (roughly the product of the current gains of the two transistors) but the base voltage drop is doubled because of the two p-n junctions in series.

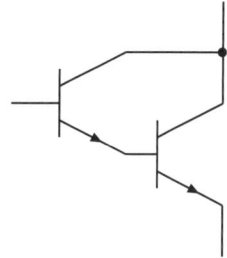

FIGURE 7.28   Darlington connection of BJTs.

7.3.2.5   Thyristors and IGBTs

The thyristor consists of four regions, doped p-n-p-n in sequence (FIGURE 7.29). When charge is injected in the gate, the barrier is lowered for injection from the anode, and the charge injected from the anode in turn lowers the barrier for charge injected from the cathode. The turn-on is regenerative, and once the device is turned on, no further charge injection is needed from the gate. The device will turn off when the applied voltage is reduced to zero, which happens in an AC circuit during zero cross over. The device can be turned off by charge extraction through the gate, but this needs a large-area gate that removes all the injected charge simultaneously, and a large gate current. Devices that are made with a turn-off gate are referred to as GTOs (gate turn-off thyristor). In the on-state, the injected charges from the anode and the cathode are equal in size and larger than the n- and p-type doping in the inner regions. This means that the device in this state behaves like a p-i-n diode, and has a low forward-voltage drop consisting of the built-in voltage of one p-n junction, and the voltage drop of the conductivity-modulated drift region. These devices require a long minority carrier lifetime to sustain the injected charge in the inner regions, but not as long as the Si counterparts since the inner regions are ten times thinner.

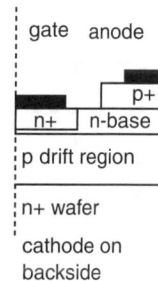

FIGURE 7.29   Thyristor (half-cell).

A device sometimes confused in operation principle with the thyristor is the insulated gate bipolar transistor, or IGBT. The equivalent circuit is that of a pnp bipolar transistor with a MOSFET gate drive. Although the device has the p-n-p-n sequence of device layers (FIGURE 7.30), the on-state operation is not that of high current injection and regenerative turn-on. The MOSFET supplies the gate current, and actually half the current is transported through the MOSFET. The charge injected in the base of the bipolar transistor is much smaller than the charge injected in the thyristor, and therefore the switching time is

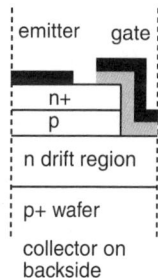

FIGURE 7.30   IGBT (half-cell).

much faster. The gate drive circuit for the IGBT does not require much power, and therefore the device is common in power circuit designs.

7.3.3 Comparison of devices

When we compare high-voltage devices the specific on-resistance is often quoted. However, this parameter is highly dependent on the blocking voltage that the device is designed for; see EQN (8) in Chapter 1. Therefore, it is better to compare the quotient between breakdown voltage squared and the on-resistance (the higher the better) as in TABLE 7.1. In FIGURE 7.31 this comparison can be done by looking at how close to the ideal on-resistance of the respective material the device is, i.e. the lines for Si, 6H

TABLE 7.1   On-resistances and breakdown voltages for some recent high-voltage devices.

Device type	Polytype	Breakdown voltage (kV)	On-resistance ($m\Omega\ cm^2$)	V_{BD}^2/R_{ON} (MW/cm^2)	Group	Ref.
pin diode	4H	19.5	65	5850	Kansai-CREE	[8,9]
pin diode	4H	4.5	42	482	RPI	[10]
JBS diode	4H	2.8	8	980	KTH-ABB	[11]
Schottky diode	4H	4.9	43	558	Purdue	[12]
DMOSFET	4H	2.4	42	137	CREE	[13]
SIAFET (cascode)	4H	6.1	732	51	Kansai-CREE	[6]
SEMOSFET (cascode)	4H	5	88	284	Kansai-CREE	[7]
ACCUFET	4H	1.4	15.7	125	Purdue	[14]
DMOSFET	6H	1.8	46	70	Siemens	[15]
JFET	4H	5.5	218	139	Kansai-CREE	[16]
VJFET	4H	2	70	57	Hitachi	[17]
VJFET	4H	3.5	25	490	SiCED	[4]
BJT	4H	1.8	10.8	300	CREE	[18]
BJT	4H	0.5	50	5	Purdue	[19]
GTO	4H	3.1	16.6	579	CREE	[20]

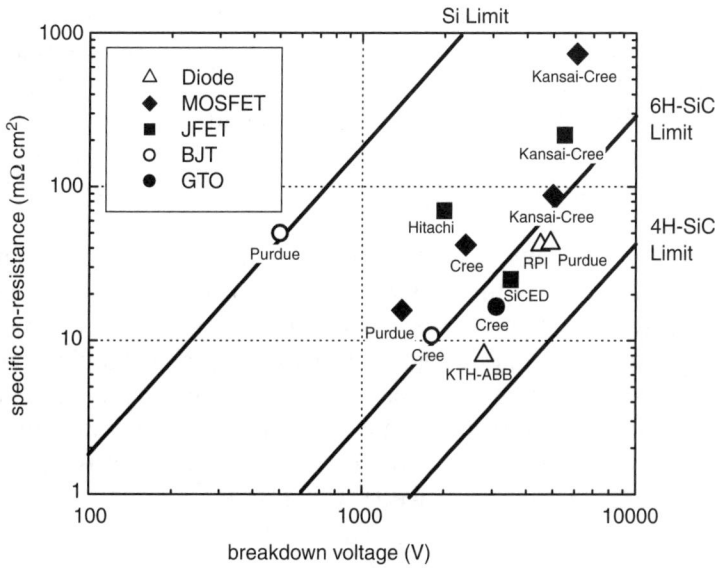

FIGURE 7.31   On-resistance versus breakdown voltage for some recent high-voltage devices.

SiC and 4H SiC. As we can see, diodes have been more successful so far, and are very close to the ideal on-resistance.

To calculate the forward-voltage drop, the on-resistance should be multiplied by the current density. A normal current density rating for a power device is 100 A/cm^2. An on-resistance of 10 mΩ cm^2 thus results in 1 V voltage drop across the drift region. For the bipolar devices there is an additional voltage drop because of the built-in voltage of the p-n junction, around 0.7 V for Si and 2.5 V for SiC. On the other hand, bipolar devices can achieve on-resistances lower than the theoretical lines when conductivity modulation is achieved, since the carrier concentration is increased by injected minority carriers. The contacts usually do not contribute any voltage drop at these current levels, since contact resistivities lower than $10^{-4}\,\Omega\,\mathrm{cm}^2$ can usually be achieved; see Chapter 6. For p-type substrates, the series resistance of the substrate can be a problem, especially at room temperature, due to incomplete ionisation. For the MOSFETs in TABLE 7.1 and FIGURE 7.31, the on-resistance is, to a large extent, the channel resistance, since the mobility of the inversion channel is still a problem; see Chapter 5.

7.3.4   Overcoming problems

For high-power devices, several devices need to be paralleled to achieve sufficient current. The load balance between the devices is very critical, since one device may be overloaded if it carries a disproportionate amount of the current. This means that the forward voltage, and hence the on-resistance, must be perfectly balanced between the devices. Subsequently, the doping variation in the drift layer is not allowed to be very large, and for bipolar devices the minority carrier lifetime must also have a small variation.

For bipolar devices, a problem has been identified recently [21]. The relatively high energy from the recombination of electrons and holes that occurs in bipolar devices in the on-state is enough to enable crystal defect growth. In-plane stacking faults, which are nucleated from screw dislocations, have been seen in several experiments, at both room temperature and elevated temperature. The effect can be modelled as a local polytype variation (a 3C SiC quantum well) or simply as a local lowering of the minority carrier lifetime. This will lead to a larger voltage drop locally, and eventually the reliability of the whole device is at stake.

For MOSFETs in SiC the problem has been seen as a low inversion mobility, 10% or less of the bulk electron mobility (in Si 50% is routinely achieved). Later, this was identified as a problem with trapping in the channel. When more than 80% of the inversion charge is fixed in the traps, the extracted channel mobility looks too small. This is further discussed in Chapter 5.

The final problem with high-voltage devices is that large areas are generally requested to carry sufficient current. Since the number of defects of various types in SiC is high (see Chapter 2) the yield for large-area devices is low. Unless the defect densities are low enough, the devices cannot be made economically. The yield Y for a device of area A, with an evenly distributed defect density D, can be calculated as

$$Y = e^{-DA} \qquad (5)$$

7.4 HIGH-FREQUENCY DEVICES

7.4.1 Introduction

This section discusses high-frequency devices in SiC, but in some cases nitride devices are included as well, since these can be major contenders, and the material can be grown on SiC substrates. After a brief description of device operation, the critical design points are covered. Some different design variations will be shown, and some different devices will be compared to each other in terms of high-frequency performance. Note that

146

the device cross sections are not drawn to scale; the vertical dimension is exaggerated.

7.4.2 Tutorial

The voltages used to bias high-frequency devices are smaller than for the high-voltage devices. Even so, all high-frequency devices benefit from the high critical field of SiC, which results in smaller and thereby faster devices. The high saturated electron velocity is also an advantage, but the mobility is a little low. However, any high-frequency device design has to take into account the parasitic elements resulting from the device layout. The speed of lateral devices is determined by all lateral dimensions, which the current has to traverse, and the capacitance of the nodes where the voltage is switched. The operating frequency is ultimately decided by the gate length of the device, and electron beam lithography is frequently used to define the gates. If the path difference is more than 1/16 of the wavelength at the operating frequency, considerable frequency dispersion will occur. The lower relative dielectric constant of SiC helps to reduce the parasitic junction capacitance, but often semi-insulating substrates are required as well.

Diodes may be used for mixers and oscillators, but the main use for high-frequency circuits is in amplifiers, where transistors are used. The comparison is split into lateral and vertical devices. Lateral devices can have small parasitics if made on semi-insulating substrates, and the speed is optimised by reducing the gate length, which places high demands on lithography. Vertical devices are perhaps slower, but may allow larger sizes for larger currents. Here, the speed is optimised in the control of epitaxial growth in terms of layer thickness and doping. Although the thermal conductivity is high, as seen in Chapter 1, it may still not be high enough, and the power limit may be set for all devices by the thermal conductivity.

7.4.2.1 Diodes

Schottky diodes are mainly used for high-frequency mixers. The devices should have very small dimensions, and are normally made without termination. A high breakdown voltage is not necessary, the advantage of SiC being the small on-resistance of the drift region. High barriers and excellent ideality factors can be achieved in SiC, which leads to low reverse leakage currents and insertion losses. To minimise the parasitic capacitance, special packages are used with whisker contacts, although the reliability has been questioned. Airbridge contacting has also been

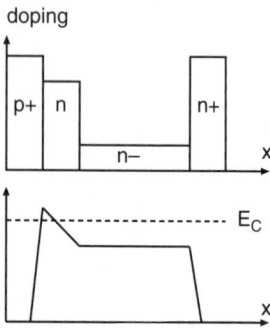

FIGURE 7.32   Doping profile and electric field variation in an IMPATT diode.

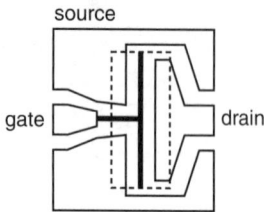

FIGURE 7.33   Lateral layout of a small high-frequency FET.

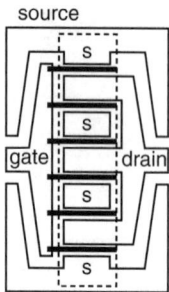

FIGURE 7.34   Lateral layout of a high-frequency FET with folded design.

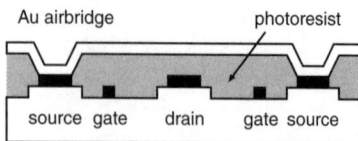

FIGURE 7.35   Airbridges are used for source interconnects.

used, and the best results have been from monolithic mixers, or where the diode has been contacted with micro-strips to the filters and matching circuit [22].

Impact ionisation avalanche transit-time (IMPATT) diodes are, unlike high-voltage diodes, designed for and used in the avalanche breakdown state. The p-n junction is between two high-doped regions, which concentrate the electric field and the carriers from impact ionisation in a small region (FIGURE 7.32). The carriers are swept away from the avalanche region, and the low-doped drift region causes a phase delay in the current signal. If this device is connected in a resonant cavity, an oscillatory signal results [23,24]. The lateral dimensions and the thickness of the doped layers are very critical for this device. The power density is very high, but the efficiency is only of the order of 1%.

7.4.2.2   Lateral devices

The lateral layout of high-frequency field-effect transistors often uses a coplanar design, and the total current of the device depends on the total gate width (FIGURE 7.33). When wider gates are needed, a folded layout is preferred, to minimise the difference in distance between different current paths (FIGURE 7.34). The contact pads are normally drawn to accommodate high-frequency probes with a GSG (ground-signal-ground) configuration for on-wafer s-parameter analysis. To connect different source parts of the device in a folded design, airbridges are used to reduce capacitance. Airbridges are made by using a disposable dielectric such as photoresist during the deposition of the metal (FIGURE 7.35). When the photoresist is removed an air gap of several micrometres results. Semi-insulating substrates are used to minimise the drain junction capacitance. To reduce the series resistance of the gates, the gate is T-shaped or mushroom shaped, to maximise the cross section while minimising gate length (FIGURE 7.36). This is accomplished by using multiple layers of photoresist. Note that the gate is normally displaced towards the drain, to optimise drain-side breakdown versus source series resistance. Finally, the source impedance from the bond wire needs to be minimised, preferably by using via holes instead. The source impedance reduces the transconductance of the transistor by negative feedback, since part of the gate voltage is dropped over this impedance, and increases with frequency. Via holes are made by etching holes through the substrate (using ICP, see Chapter 4), after thinning the substrate to 50–100 μm, and then electroplating the holes and backside with Au (FIGURE 7.37).

The metal-semiconductor field-effect transistor (MESFET) uses a Schottky metal gate to deplete the channel between source and drain. The depletion region from the reverse-biased gate is controlled with the voltage, and these are normally-on devices. The maximum current is determined by the channel dose (doping and depth) and is a compromise regarding the pinch-off voltage. There have been problems with charge trapping in the p-type buffer layer between channel and semi-insulating substrate, which changes the pinch-off voltage and causes high-frequency properties to shift.

The junction field-effect transistor (JFET) with a similar operating principle is seldom made for high-frequency use, since the p-n junction gate is difficult to make as short as a metal gate. The metal-oxide-semiconductor field-effect transistor (MOSFET) can be used as a depletion mode device (see discussion in Section 7.3.2.3) for high-frequency use. These are normally-on devices, with similar design and mobility to those of the MESFET, but the insulating gate may be forward biased, unlike the Schottky gate, which allows larger currents. However, the gate capacitance can be difficult to match in a circuit. Most common is the LDMOS-FET for high-frequency design [25] (FIGURE 7.38). It is important that the polysilicon gate is metallised to reduce the series resistance.

Finally, we need to include the high electron mobility transistor (HEMT), also called the modulation doped field-effect transistor (MODFET), in the discussion. This device uses epitaxial layers of AlGaN on GaN, two wide bandgap semiconductors (FIGURE 7.39). These materials can be grown on SiC wafers, but may also use sapphire. The reason for the comparison is that this device is the main contender for the SiC MESFET for high-frequency high-power devices. AlGaN has a wider bandgap than GaN, and a 2D electron gas channel is induced between the two materials. The charge in this channel can be very high, and is controlled using a Schottky gate on the surface. Since the charge is collected in an unintentionally doped region of GaN, the mobility is very high, and hence the name HEMT. Passivation of the surface is very important, as well as the material growth, since polarisation charges as large as the channel charge can occur in these materials [26].

7.4.2.3 Vertical devices

The speed of vertical devices is determined by all vertical dimensions, which the current has to traverse, and the capacitance of the nodes where the voltage is switched. The operating frequency is ultimately decided by the base width of the device,

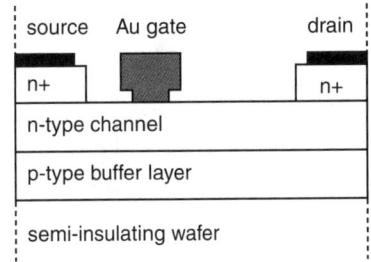

FIGURE 7.36 High-frequency MESFET with recessed T-gate.

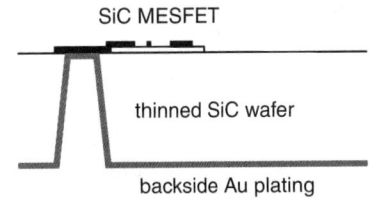

FIGURE 7.37 Via holes are used to reduce the source impedance.

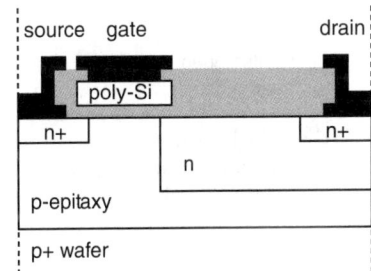

FIGURE 7.38 High-frequency LDMOSFET (half-cell).

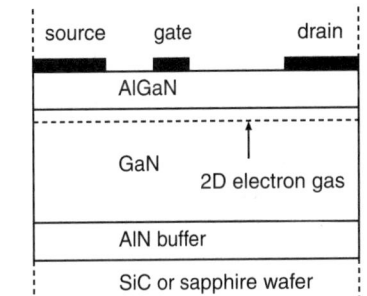

FIGURE 7.39 AlGaN/GaN HEMT on SiC.

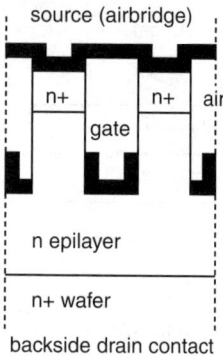

FIGURE 7.40   Static induction transistor (SIT) (two cells).

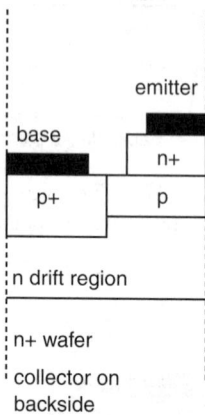

FIGURE 7.41   Bipolar high-frequency transistor with epitaxial emitter and ion-implanted base (half-cell).

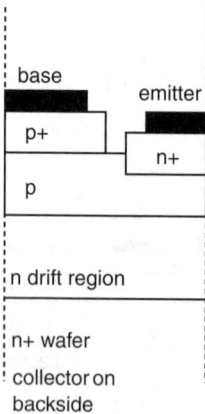

FIGURE 7.42   Bipolar high-frequency transistor with implanted emitter and etched base (half-cell).

which sets high demands on the epitaxial growth of thin layers with well-defined doping. The lateral size is limited by the wavelength at operation, since the path difference has to be minimised just as in the lateral devices, and therefore the lateral layout is similar. However, since one contact (normally the collector) is placed on the back of the substrate, in principle more base area should be possible and therefore higher currents. Often the collector is contacted from the front as well, and then semi-insulating wafers are preferred.

The static induction transistor (SIT), also known as the permeable base transistor (PBT), can be compared in operation to a vertical field-effect transistor or a vacuum triode. The gate control can use either a p-n junction or a Schottky metal, to deplete the channel regions between the gates (FIGURE 7.40). Several channels are made in parallel by reactive ion etching, and these need to be narrow enough (submicrometre) that they can be pinched off. Airbridges are needed to connect the sources.

The bipolar junction transistor (BJT) was discussed extensively in Section 7.3.2.4. The high-frequency version has a narrow base for higher frequency (shorter transport length) and the operational voltage is therefore lower. All designs have a compromise between high base doping (achieves better base contacts, lowers base resistance for higher f_{max}, and avoids base punch-through) and high gain (needs lower doping in the intrinsic base and regrowth or ion implanting of the extrinsic base). The Kirk effect is negligible due to higher collector doping in SiC. If an epitaxial emitter is used, the etching in the patterning of the emitter needs to stop at the base, which puts high demands on the etching process. Usually this needs to be combined with ion-implanted base extensions or regrowth (FIGURE 7.41). If an implanted emitter is used, a thicker base layer can be used, since the base width is determined by the ion-implanted junction depth after the base has been etched (FIGURE 7.42). The disadvantage of ion-implanted emitters is that the remnant damage in the base region below the emitter lowers the minority carrier lifetime and reduces the gain.

A solution to the compromise in the base doping was also mentioned in Section 7.3.2.4, the heterojunction bipolar transistor (HBT). By using an emitter with a wider bandgap, a high base doping can be allowed without sacrificing the current gain (FIGURE 7.43). The difficulty is to grow GaN or AlGaN on SiC. The lattice mismatch causes large numbers of traps in the interface, which reduces injection of carriers from the emitter. The type II bandgap lineup possibly enhances recombination as well, which may be the reason for few successful results so far [27]. For an easy process, selective growth of the nitride material would be preferred, even if highly selective etching of nitrides on SiC can be achieved.

7.4.3   Comparison of devices

There are several figures of merit for comparing high-frequency devices, such as maximum operating frequency (f_T) or maximum frequency of oscillation (f_{max}). However, to be able to compare smaller devices with larger devices, it is usual to scale the output power with the gate width, and do the comparison at the operating frequency. Any device used in a circuit to measure the output power needs an f_T and f_{max} about ten times higher than the operating frequency, as a rule of thumb. Since there are different device applications, devices in the lower left corner are

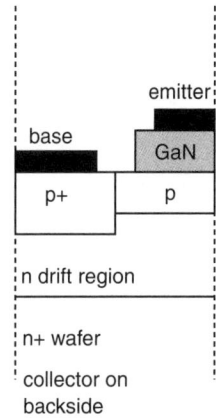

FIGURE 7.43   Heterojunction bipolar transistor (half-cell).

TABLE 7.2   Output power density and operating frequency for some recent high-frequency devices.

Device type	Polytype	Operating frequency (GHz)	Power density (W/mm)	Measured mode	Total power (W)	Group	Ref.
IMPATT diode	4H	9.9	300 mW	Pulsed		Ioffe Inst.	[23]
IMPATT diode	4H	7.2	12 dBm 400 mA 9 kA/cm^2	Pulsed		Purdue	[24]
MESFET	4H	3.5	5.2	Pulsed	62	CREE	[28,29]
MESFET	4H	10	4.5	Pulsed	54	CREE	[29]
MESFET	4H	2	4	CW	8	Thomson	[30]
SIT	4H	1.3	1.67	Pulsed	400	Northrop Grumman	[31]
SIT	4H	2.9	1.51	Pulsed	78	Northrop Grumman	[32]
GaN HEMT		3.5	12.1	Pulsed	145	Cree	[28]
AlGaN HEMT		10	4.2	Pulsed	50.1	Cree	[28]
AlGaN HEMT		20	6.6	CW	1.32	HRL Labs.	[33]
Si MOSFET		1.9	0.32	CW	2.4	Hitachi	[34]
GaAs MODFET		2.16	2.22	Pulsed	200	Matsushita	[35]

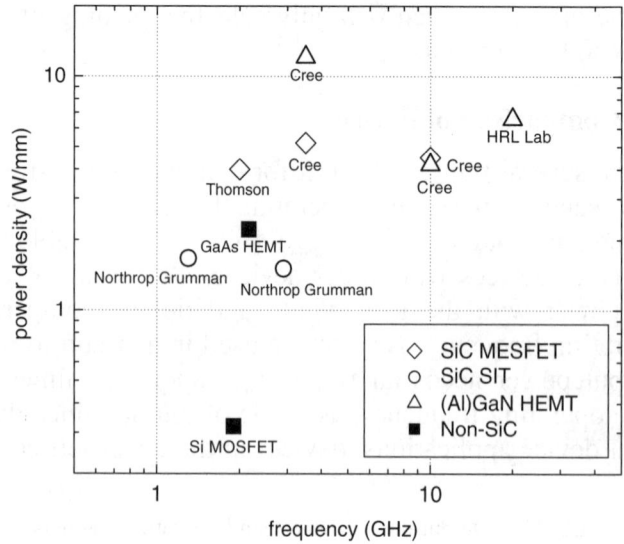

FIGURE 7.44   Output power density versus operating frequency for some recent high-frequency devices.

certainly not worthless, especially if the price is much lower. Most of the promising results have been achieved using lateral devices.

To achieve sufficient power, wider gates, i.e. larger devices, can be used to a certain limit, which is when the device size is roughly 1/16 the size of the operational wavelength. Therefore, higher power per unit gate width is of interest for high-frequency devices. However, the cooling requirements of such devices can be daunting at these power densities. Although the thinning of the substrates for the via holes does improve the thermal conductivity somewhat, it is not expected that the thermal resistance of the solder joint between substrate and carrier can be much better than in earlier generations of devices.

7.4.4   Overcoming problems

For precise control of the device properties of high-frequency devices, good control is needed of the epitaxial growth process in terms of doping and thickness control. If regrowth can be used for highly doped base contacts, better devices can be made.

For lateral nitride devices, heteroepitaxial growth is needed since nitride substrates are still not available in large quantities. SiC is the better substrate in terms of thermal conductivity. Some special growth processes have been investigated, such as lateral epitaxial overgrowth (LEO or sometimes ELO) or pendeoepitaxy [36], where oxide masks are used for selective growth in smaller areas. Although this works for lateral devices, for vertical devices the interface is still problematic, and other techniques are needed.

Etching is another key process. The recessed channel needs to be etched with precise thickness control and without remnant damage to avoid excessive gate leakage. Etching the emitter on top of a very thin base is also challenging, but photoelectro-chemical etching should be investigated for this purpose, since it has shown some selectivity between doping types. Via holes need very high etch rates, since 100-μm deep holes are needed even after the substrates are thinned.

7.5 HIGH-TEMPERATURE, OPTICAL AND MECHANICAL DEVICES

7.5.1 Introduction

This section covers the devices that cannot simply be classified as high-voltage, or high-frequency devices, but still take advantage of some material property of SiC. These devices are included for the sake of completeness, as they also are made using the process technology described in this book. After a brief discussion about device operation, some device examples will be shown. Note that the device cross sections are not drawn to scale; the vertical dimension is exaggerated.

7.5.2 Tutorial

7.5.2.1 High-temperature devices

The wide bandgap of SiC results in very low intrinsic carrier concentration, which means that SiC devices could theoretically operate up to 1000 °C. This would allow the application of electronics in high-temperature environments, such as in or close to combustion engines and other hot processes. Although both digital and analogue circuits have been demonstrated using MOSFETs and JFETs, respectively, this does not seem to be a big market yet. One problem is the long-term reliability of devices and circuits, which will probably limit operation before the material itself. Packaging of high-temperature electronics is another area in need of research. Finally, threshold voltages and operating points will shift with temperature, so the electronic circuit design will need to take this into account.

An application that seems closer to actual use is catalytic gas sensors. The basic principle is to use a catalytic metal (Pt, Pa, Ir etc.) as a gate on a metal-oxide-semiconductor (MOS) structure

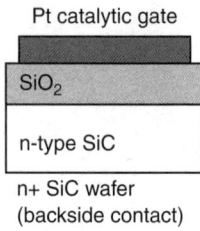

FIGURE 7.45 High-temperature gas sensor with catalytic gate.

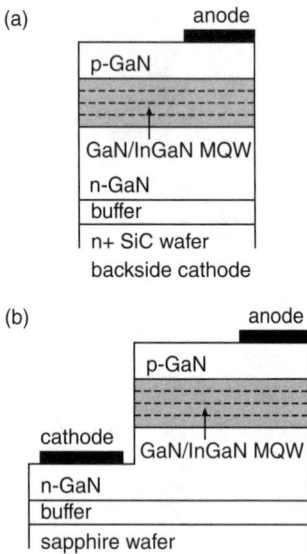

FIGURE 7.46 (a) GaN MQW LED on SiC substrate;
(b) GaN MQW LED on sapphire.

(FIGURE 7.45). Some gases will crack or decompose on the catalytic metal, and hydrogen ions can be sensed on the gate by the shift in flatband voltage that the extra charge produces. Several devices based on this principle have been demonstrated in several different sensor applications for combustion processes [37].

Temperature sensors for higher temperatures are another possibility. Almost any electronic device can be used as a temperature sensor; it is mainly a question of finding a property that can be measured accurately in the temperature range, and which preferably has a linear temperature dependence. Reverse leakage current in diodes is one alternative. Using the change in a transistor parameter, such as the current gain, adds to the sensitivity if the amplification in the device can be used.

7.5.2.2 Optical devices

Photodetectors may be made in SiC, if sensitivity in or above the UV range is needed. Since SiC is an indirect semiconductor, these devices may also be investigated using GaN, which has a similar but direct bandgap. Blue LEDs were initially made in SiC, but with very low efficiency. Once the LEDs could be made by GaN/InGaN multiple quantum wells (MQW) this effort was ended. However, SiC is the preferred substrate for GaN LEDs or lasers. The advantage over sapphire is not only the higher thermal conductivity, but also that the SiC substrates conduct current. Therefore, LEDs can be made with one frontside and one backside contact with SiC substrates (FIGURE 7.46(a)), but with sapphire substrates two frontside contacts and less conventional packages have to be used (FIGURE 7.46(b)). These devices require that a conductive buffer can be grown on SiC.

7.5.2.3 Mechanical devices (MEMS)

The final type of device takes advantage of the mechanical properties of SiC. Microelectromechanical systems (MEMS) have been pursued in Si technology for a long time [38]. Some applications are diaphragms as pressure sensors, cantilevers for acceleration sensors, movable mirrors for optical switching, etc. The advantage of SiC is that it is more rugged than Si, and has a higher mechanical velocity due to the harder chemical bond. Usually, the electrical sensing is done through metal resistors used as strain gauges, which means that the SiC can be polycrystalline. Since polycrystalline 3C SiC can be grown on Si wafers, this opens up very large markets, since larger wafers and standard fabs can be used. The main process step for MEMS is micromachining, which is done using ICP for Si as well. An advantage with poly-SiC on Si is that the Si material can be removed selectively using wet etching, so that free-standing membranes can be made; see FIGURE 7.47. For high-pressure sensing of corrosive gases, the metal resistors can be made on the protected side of the membrane. Radio frequency (RF) filters using surface acoustic waves (SAW) is another possibility, many times further enhancing the response with AlN films grown on SiC.

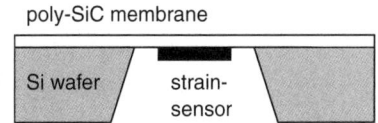

FIGURE 7.47   SiC pressure sensor on Si wafer.

7.5.3   Overcoming problems

Long-term stability is an issue for any device to be used in extreme environments. The cost has been discussed, but in some areas like catalytic gas sensors, where few alternatives exist, price is not an issue. For MEMS devices, large-area substrates exist. This is probably an area where the imagination is the limit.

7.6   RECENT DEVELOPMENTS AND FUTURE TRENDS

Although the high critical field of SiC promises high-voltage devices with much lower on-resistance than Si devices, this has been difficult to achieve in practice. The defect problems, in general, make large-area devices difficult to fabricate, and especially the bipolar defect [21] could be a showstopper for these devices. Although recent progress has been made in terms of inversion channel mobility for MOSFETs, the highest breakdown voltages need bipolar devices for low on-resistance through conductivity modulation.

Most high-frequency work today is concentrated on improving the growth of nitrides, since HEMTs have shown much better high-frequency characteristics than MESFETs. SiC would then be an ideal substrate in the absence of nitride substrates, due to better thermal conductivity than sapphire. If HBTs can be made, vertical bipolar devices may also be contenders in this field.

High-temperature gas sensors for monitoring combustion pressures show large promise as environmental concerns increase. For optical applications, SiC has been demoted to substrate material, but only until nitride substrates can be made. MEMS is an area where SiC can be processed on substrates in large dimensions, but this really only uses polycrystalline SiC.

7.7   CONCLUSIONS

Even though the basic material properties discussed in Chapter 1 would seem to promise vastly improved devices, there are many difficulties to be solved. Once the process technologies that are needed have been developed (including growth, doping, etching, isolation and contacts discussed in Chapters 2–6) the process steps need to be put together. It is not until completed devices can be measured in realistic circuit configurations that the comparison can be fairly made to Si and GaAs counterparts. Even superior devices may not prevail in the end, due to questions of economy resulting partly from yield. This chapter has shown how to put together the different process steps, and some common device structures that have been made so far. Yet other device structures may be suggested that can better utilise the advantageous semiconductor properties of SiC.

REFERENCES

[1] S.M. Sze [*Physics of Semiconductor Devices* 2nd edn (Wiley-Interscience, 1981)]

[2] B.J. Baliga [*Power Semiconductor Devices* (PWS Publishing Company, 1996)]

[3] S.-M. Koo, S.-K. Lee, C.-M. Zetterling, M. Ostling, U. Forsberg, E. Janzén [*Mater. Sci. Forum (Switzerland)* vol.389–393 (2002) p.1235–8]

[4] P. Friedrichs, H. Mitlehner, R. Schörner, K.-O. Dohnke, R. Elpelt, D. Stephani [*Mater. Sci. Forum (Switzerland)* vol.389–393 (2002) p.1185–90]

[5] J.A. Cooper Jr., M.R. Melloch, R. Singh, A. Agarwal, J.W. Palmour [*IEEE Trans. Electron Devices (USA)* vol.49 (2002) p.658–64]

[6] D. Takayama et al [*IEEE International Symposium on Power Semiconductor Devices and ICs (ISPSD) (USA)* (2001) p.41–4]

[7] Y. Sugawara et al [*Mater. Sci. Forum (Switzerland)* vol.389–393 (2002) p.1199–202]

[8] CREE Inc. Durham, NC, USA, http://www.cree.com

[9] Y. Sugawara, D. Takayama, K. Asano, R. Singh, J. Palmour, T. Hayashi [*IEEE International Symposium on Power Semiconductor Devices and ICs. ISPSD '01 (USA)* (2002) p.27–30]

[10] J.B. Fedison, N. Ramungul, T.P. Chow, M. Ghezzo, J.W. Kretchmer [*IEEE Electron Device Lett. (USA)* vol.22 (2001) p.130–2]

[11] F. Dahlquist, J.O. Svedberg, C.M. Zetterling, M. Ostling, B. Breitholtz, H. Lendenmann [*Mater. Sci. Forum (Switzerland)* vol.338 (2000) p.1179–82]

[12] H.M. McGlothlin, D.T. Morisette, J.A. Cooper Jr., M.R. Melloch [*57th Annual Device Research Conference Digest (USA)* (1999) p.42–3]

[13] S.H. Ryu, A. Agarwal, M. Das, L. Lipkin, J. Palmour, N. Saks [*Mater. Sci. Forum (Switzerland)* vol.389–393 (2002) p.1195–8]

[14] J. Tan, J.A. Cooper Jr., M.R. Melloch [*IEEE Electron Device Lett. (USA)* vol.19 (1998) p.487–9]

[15] D. Peters, R. Schorner, P. Friedrichs, J. Volkl, H. Mitlehner, D. Stephani [*IEEE Trans. Electron Devices (USA)* vol.46 (1999) p.542–5]

[16] K. Asano et al [*IEEE International Symposium on Power Semiconductor Devices and ICs (ISPSD) (USA)* (2001) p.23–6]

[17] H. Onose, A. Watanabe, T. Someya, Y. Kobayashi [*Mater. Sci. Forum (Switzerland)* vol.389–393 (2002) p.1227–30]

[18] S. Ryu, A.K. Agarwal, R. Singh, J.W. Palmour [*IEEE Electron Device Lett. (USA)* vol.22 (2001) p.124–6]

[19] Y. Tang, J.B. Fedison, T.P. Chow [*IEEE Electron Device Lett. (USA)* vol.23 (2002) p.16–8]

[20] S.-H. Ryu, A.K. Agarwal, R. Singh, J.W. Palmour [*IEEE Electron Device Lett. (USA)* vol.22 (2001) p.127–9]

[21] J.P. Bergman, H. Lendenmann, P.Å. Nilsson, U. Lindefelt, P. Skytt [*Mater. Sci. Forum (Switzerland)* vol.353–356 (2001) p.299–302]

[22] J. Eriksson, N. Rorsman, F. Ferdos, H. Zirath [*Electron. Lett. (UK)* vol.37 (2001) p.54–5]

[23] K.V. Vassilevski et al [*Mater. Sci. Forum (Switzerland)* vol.389–393 (2002) p.1353–8]

[24] L. Yuan, J.A. Cooper, K.J. Webb, M.R. Melloch [*Mater. Sci. Forum (Switzerland)* vol.389–393 (2002) p.1359–62]

[25] D. Alok et al [*IEEE Electron Device Lett. (USA)* vol.22 (2001) p.577–8]

[26] S.C. Binari et al [*IEEE Trans. Electron Devices (USA)* vol.48 (2001) p.465–71]

[27] E. Danielsson, C.-M. Zetterling, M. Ostling, D. Tsvetkov, V.A. Dmitriev [*J. Appl. Phys. (USA)* vol.91 (2002) p.2372–9]

[28] J.W. Palmour et al [*Int. Electron Devices Meet. Tech. Dig. (USA)* (2001) p.17.14.11–14]

[29] R.A. Sadler, S.T. Allen, W.L. Pribble, T.S. Alcorn, J.J. Sumakeris, J.W. Palmour [*IEEE Cornell Conference on High Performance Devices* Piscataway, NJ, USA (2000) p.173–7]

[30] O. Noblanc, C. Arnodo, C. Dua, E. Chartier, C. Brylinski [*Mater. Sci. Forum (Switzerland)* vol.338 (2000) p.1247–50]

[31] R.C. Clarke, A.W. Morse, P. Esker, W.R. Curtice [*IEEE Cornell Conference on High Performance Devices* Piscataway, NJ, USA (2000) p.141–3]

[32] R.C. Clarke et al [*IEEE High-Temperature Electronic Materials, Devices and Sensors Conference* New York, NY, USA (1998) p.96–7]

[33] J.S. Moon et al [*Electron. Lett. (UK)* vol.37 (2001) p.528–30]

[34] I. Yoshida, M. Katsueda, Y. Maruyama, I. Kohjiro [*IEEE Trans. Electron Devices (USA)* vol.45 (1998) p.953–6]

[35] H. Ishida et al [*Int. Electron Devices Meet. Tech. Dig. (USA)* (1999) p.393–6]

[36] R.F. Davis et al [*J. Cryst. Growth (Netherlands)* vol.225 (2001) p.134–40]

[37] A. Lloyd Spetz et al [*Mater. Sci. Forum (Switzerland)* vol.389–393 (2002) p.1415–8]

[38] T.-R. Hsu [*MEMS & Microsystems: Design and Manufacture* (McGraw-Hill, 2002)]

Appendix 1: Glossary

These definitions are given within the framework of this book and are mainly applicable to silicon or silicon carbide device processing. Items in *italics* can be found in this glossary, whereas more details are in the referenced chapters.

Alloying

The process of mixing metals, or reacting metals with SiC, which is often done at elevated temperature in a controlled atmosphere. The purpose is to improve the *Schottky* or *ohmic contacts*: see Chapter 6.

Bipolar device

Electronic device relying on electron and hole transport for its operation. The injection of minority carriers is of importance, as it lowers the on-resistance through conductivity modulation. Examples of bipolar devices are p-i-n diodes, bipolar junction transistors (BJTs) and thyristors: see Chapter 7.

Bulk growth

The process of growing single-crystal SiC boules so that entire wafers can be cut from them. This process in SiC is characterised by very high temperatures, above 2300 °C: see Chapter 2.

Cascode circuit

The cascode circuit is the combination of two transistors, where one is connected with its source or emitter to the output (the collector or drain terminal) of the other: see Chapter 7. The purpose is to change the output conductance of the pair, or distribute the high voltage so that one device blocks most of the voltage.

Deposition

This refers to any process where the material that is deposited is added; none of it is taken from the wafer. The opposite is, for example, *thermal oxidation* or *silicidation*, which consumes some SiC. The precursors can be in the gas phase (*epitaxial growth*), liquid phase (*evaporation*) or solid phase (*sputtering*): see Chapters 2 and 6.

Diffusion

This is the general process of impurity redistribution at higher temperatures, as impurities will tend to distribute evenly in the semiconductor. Doping by diffusion of dopants into the SiC wafer requires such high temperatures (more than 1800 °C) that masking is difficult. Doping by diffusion is very seldom used in SiC process technology: see Chapter 2.

Dry etching

An etching procedure where a gas plasma of ionised etchants is used to remove material, either by forming volatile species or by sputtering. Examples of equipment types for dry etching are reactive ion etching (RIE), electron cyclotron resonance (ECR) and inductively coupled plasma (ICP): see Chapter 4.

Epitaxial growth

Growth of a material using the underlying material as a template for the growing crystal. This is only possible if the lattice constant is almost the same for the two materials. Epitaxial growth of SiC on SiC wafers is used to achieve the small doping concentrations needed to block high voltages, and to vary the doping. Lateral patterning with *dry etching* is often used in combination with epitaxial growth. Two common processes are chemical vapour deposition (CVD) and vapour phase epitaxy (VPE): see Chapter 2.

Evaporation

If the temperature is high enough, or the pressure low enough, any material will evaporate. To enhance the process for deposition of metals through evaporation, the metal is heated either thermally or with a focused electron beam inside a vacuum chamber. The evaporated material condenses on the wafer: see Chapter 6.

Field-effect transistor (FET)

Electronic switching device where an electric field from the voltage applied to a gate is used to control the current from the source contact to the drain contact. These are *unipolar devices*: see Chapter 7.

Heterojunction

The junction between two different semiconductors, often with different energy bandgaps. Normally all p-n junctions in a device are homojunctions, as the device is made entirely in Si or SiC. By joining two different semiconductors the band structure can be tailored. Examples of devices using heterojunctions are HBTs and HEMTs (sometimes called MODFETs): see Chapter 7.

Ion implantation

This is a process for doping of a semiconductor by accelerating dopant ions in an electric field, and bombarding the wafer with the ions. By masking part of the surface, lateral variation can be achieved. Ion implantation allows very good dose and depth control, but requires a high-temperature anneal afterwards to remove the damage caused by the ions. For SiC, it is also preferred to have a high temperature during ion implantation: see Chapter 3.

Lift-off

This is a process for patterning metals without *wet etching*, which is useful if the metal used is difficult to etch. Before the metal is *deposited*, photoresist is deposited and patterned. After metal deposition the portion of the metal on top of the remaining photoresist can be removed by dissolving the photoresist with acetone: see Chapter 6.

Ohmic contact

A contact between a metal and SiC that has no barrier for current transport, and hence conducts current equally well in forward and reverse bias; see Chapter 6. To achieve an ohmic contact, high doping ($>10^{19}$ cm^{-3}) and high annealing temperatures (for *alloying* or *silicidation*) are needed. A low *Schottky barrier* between the metal and SiC is also preferred.

Passivation

If the surface of SiC is not passivated, there can be charges that change the electric field inside the device. This can cause premature breakdown or threshold voltage shift. One common method of passivation is *thermal oxidation* of the SiC surface: see Chapter 5.

Regrowth

If *epitaxial growth* of SiC is performed after parts of the surface have been modified through *dry etching* or *ion implantation*, this is called regrowth. The equipment involved is the same as for *epitaxial growth*: see Chapter 2.

Sacrificial oxidation

This is the process of *thermally oxidising* the SiC (thereby consuming Si from the SiC), and then removing the grown oxide: see Chapter 5. This is a controlled way of removing SiC without causing damage, contrary to *dry etching*. Often, this is performed after *dry etching* or *ion implantation* to remove damaged SiC.

Schottky barrier

The barrier for electrons and holes to flow from the metal to the semiconductor. The magnitude of the Schottky barrier is determined by the metal work function. Current transport across a Schottky barrier is lower in reverse bias compared to forward bias: see Chapter 6. This type of contact is commonly achieved if the doping in the SiC is low ($<10^{17} \mathrm{cm}^{-3}$), and is used in Schottky diodes and MESFETs: see Chapter 7.

Silicidation

The process of forming a silicide (a compound consisting of a metal and Si). Unless Si has been deposited, it will be consumed from the SiC wafer, at the same time freeing C. Carbides (compounds consisting of a metal and C) may be formed at the same time. High temperatures are required, and commonly a controlled atmosphere is used. Used for *ohmic contacts*: see Chapter 6.

Specific on-resistance

Also referred to as resistivity, but specific on-resistance is preferred. This is the resistance per surface area in $\Omega\,\mathrm{cm}^2$, and is commonly used to characterise *ohmic contacts*: see Chapter 6. To calculate the voltage drop in a contact, multiply by the current density in A/cm².

Sputtering

The process of sputtering, in general, refers to removal of a material through momentum transfer from energetic ions. Sputter deposition of metals uses a plasma to sputter a metal or alloy from a target, and the sputtered material is then redeposited on the wafer. The advantages of this process over *evaporation* are good step-coverage, and that compound materials can be deposited: see Chapter 6.

Termination

When a p-n junction is reverse biased, electric field crowding can occur if one side of the p-n junction has a much smaller area. Premature breakdown can be avoided through termination of the electric field over a larger area, and special termination structures are used to spread the electric field: see Chapter 7.

Thermal oxidation

This is the process of growing silicon dioxide (SiO_2) on SiC. In contrast to oxide *deposition*, Si is consumed from the SiC wafer, and the process has to be done at elevated temperature in an atmosphere containing pure oxygen or water vapour: see Chapter 5.

Unipolar device

Electronic device that only needs one type of carrier for current transport, usually electrons since the electron mobility is higher. Examples of devices are Schottky diodes and all types of field-effect transistors (JFETs, MESFETs, MOSFETs, HEMTs, etc.): see Chapter 7.

Wet etching

An etching procedure where the etchant is in liquid phase or dissolved in a solution, and the material is submerged in the etchant. Mainly used for patterning metals, as SiC is very difficult to wet etch: see Chapter 4.

Appendix 2: Other resources

Conferences

The main conference for the SiC researchers is the International Conference on Silicon Carbide and Related Materials (ICSCRM), which is biannual. In between these conferences, which rotate between Europe, USA and Japan, there is a European counterpart called ECSCRM. The latest conference proceedings from these two conference series are available from TTP publishers as Materials Science Forum, ISSN: 0255-5476 (www.ttp.net) [1,2]. The Materials Research Society (MRS) has also had several symposia on silicon carbide and other wide bandgap materials at its Spring and Fall meetings starting from around 1994. Symposia proceedings are available from MRS (www.mrs.org) and the most recent volumes are numbers 339, 423, 483, 512, 572, 622 and 640.

Other general references

For more detailed references on some of the material properties in Chapter 1, see EMIS Datareviews Series No. 13, Properties of Silicon Carbide [3]. Two other general references, which cover both specific processing and device topics in SiC, are also available [4,5]. There are many more books written about Si process technology and devices, but two that I can recommend are [6] and [7].

Companies and research groups

SiC is researched worldwide, and there are several companies and research groups in Europe, USA and Japan. The companies which manufacture SiC wafers or SiC process equipment are most easily found by a search on the Internet. The research groups are found in universities, government agencies and companies, and their results are published in many international journals and the above-mentioned conference proceedings. Some groups may have resources available on the Internet as well. However, including a list of web addresses here is difficult, since it would be impossible to ensure that it was accurate and complete.

REFERENCES

[1] S. Yoshida, S. Nishino, H. Harima, T. Kimoto (Eds.) [*Proc. 9th International Conference on Silicon Carbide and Related Materials (ICSCRM2001), Tsukuba, Japan, 28 Oct.–2 Nov. 2001, Mater. Sci. Forum (Switzerland)* vol. 389–393 (Trans Tech Publications Inc., Zurich, 2002)]

[2] G. Pensl, D. Stephani, M. Hundhausen (Eds.) [*Proc. 3rd European Conf. on Silicon Carbide and Related Materials (ECSCRM2000), Kloster Banz, Germany, 3–7 Sept. 2000 Mater. Sci. Forum, (Switzerland)* vol. 353–356 (Trans Tech Publications Inc., Zurich, 2001)]

[3] G.L Harris (Ed.) [*Properties of Silicon Carbide,* EMIS Datareviews Series No. 13 (INSPEC, IEE, London, UK, 1995)]

[4] W.J. Choyke, H. Matsunami, G. Pensl (Eds.) [*Silicon Carbide, A Review of Fundamental Questions and of Applications to Current Device Technology* (Wiley, 1997) or (Akademie Verlag, Berlin, 1997)]

[5] Yoon Soo Park (Ed.) [*SiC Materials and Devices, Semicond. Semimet.* vol.52 (Academic Press, New York, 1998)]

[6] J.D. Plummer, M.D. Deal, P.B. Griffin [*Silicon VLSI Technology* (Prentice Hall, 2000)]

[7] S.M. Sze [*Physics of Semiconductor Devices* 2nd edn (Wiley-Interscience, 1981)]

Subject index